Grade 4

21st Century Assessments

www.mheonline.com

Copyright ©2015 McGraw-Hill Education

All rights reserved. The contents, or parts thereof, may be reproduced in print form for non-profit educational use with *McGraw-Hill My Math*, provided such reproductions bear copyright notice, but may not be reproduced in any form for any other purpose without the prior written consent of McGraw-Hill Education, including, but not limited to, network storage or transmission, or broadcast for distance learning.

 McGraw-Hill is committed to providing Instruction materials in Science, Technology, Engineering, and Mathematics (STEM) that give all students a solid foundation, one that prepares them for college and careers in the 21st century.

Send all inquiries to:
McGraw-Hill Education
8787 Orion Place
Columbus, OH 43240

ISBN: 978-0-07-667439-8
MHID: 0-07-667439-8

Printed in the United States of America.

1 2 3 4 5 6 7 8 9 10 RHR 20 19 18 17 16 15

 Our mission is to provide educational resources that enable students to become the problem solvers of the 21st century and inspire them to explore careers within Science, Technology, Engineering, and Mathematics (STEM) related fields

Contents

Teacher's Guide
How to Use this Book iv

Assessment Item Types
Selected-Response Items vii
Constructed-Response Items xi

Countdown
Weeks 20 to 1 13

Chapter Tests
Chapter 1 Test 53
Chapter 2 Test 59
Chapter 3 Test 65
Chapter 4 Test 71
Chapter 5 Test 77
Chapter 6 Test 83
Chapter 7 Test 89
Chapter 8 Test 95
Chapter 9 Test 101
Chapter 10 Test 107
Chapter 11 Test 113
Chapter 12 Test 119
Chapter 13 Test 125
Chapter 14 Test 131

Chapter Performance Tasks
Chapter 1 Performance Task 137
Chapter 2 Performance Task 139
Chapter 3 Performance Task 141
Chapter 4 Performance Task 143
Chapter 5 Performance Task 145
Chapter 6 Performance Task 147
Chapter 7 Performance Task 149
Chapter 8 Performance Task 151
Chapter 9 Performance Task 153
Chapter 10 Performance Task 155
Chapter 11 Performance Task 157
Chapter 12 Performance Task 159
Chapter 13 Performance Task 161
Chapter 14 Performance Task 163

Benchmark Tests
Benchmark Test, Chapters 1–5 165
Benchmark Test, Chapters 6–10 173
Benchmark Test A, Chapters 1–14 181
Benchmark Test B, Chapters 1–14 193

Answers and Rubrics
Countdown 209
Chapter Tests 229
Performance Task Rubrics
 and Sample Student Work 271
Benchmark Tests 355

Teacher's Guide to 21st Century Assessment Preparation

Whether it is the print *21st Century Assessments* or online at **ConnectED**. mcgraw-hill.com, *McGraw-Hill My Math* helps students prepare for online testing.

How to Use this Book

21st Century Assessments includes experiences needed to prepare students for the upcoming online state assessments. The exercises in this book give students a taste of the different types of questions that may appear on these tests.

Assessment Item Types

- Familiarizes students with commonly-seen item types
- Each type comes with a description of the online experience, helpful, hints, and a problem for students to try on their own.

Countdown

- Prepares students in the 20 weeks leading up to the online assessments
- Consists of five problems per week, paced with order of the *McGraw-Hill My Math Student Edition* with built in review.
- **Ideas for Use** Begin use in October for pacing up to the beginning of March. Assign each weekly countdown as in-class work for small groups, homework, a practice assessment, or a weekly quiz. You may assign one problem per day or have students complete all five problems at once.

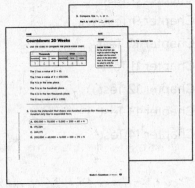

Chapter Tests

- Each six-page test assesses all of the standards for mathematics presented in the chapter.
- Each question mirrors an item type that might be found on online assessments, including multi-part questions.
- **Ideas for Use** Assign as in-class group work, homework, a practice assessment, a diagnostic assessment before beginning the chapter, or a summative assessment upon completing the chapter.

Chapter Performance Tasks

- Each two-page performance task measures students' abilities to integrate knowledge and skills across multiple standards. This helps students prepare for the rigor expected in college and future careers.

- A rubric describes the standards assessed and guidelines for scoring student work for full and partial credit.

- Sample student work is also included in the answer section of this book.

- **Ideas for Use** Assign as in-class small group work, homework, a practice assessment, or in conjunction with the Chapter Test as part of the summative assessment upon completion of the chapter.

Benchmark Tests

Four benchmark tests are included in this book. All problems on the tests mirror the item types that may be found on online assessments. Each benchmark test also includes a performance task.

- The *first* benchmark test is an eight-page assessment that addresses the standards from the first third of the Student Edition.

- The *second* benchmark test addresses the second third of the Student Edition.

- The *third* and *fourth* benchmark tests (Forms A and B) are twelve-page assessments that address the standards from the entire year, all chapters of the Student Edition.

- A rubric is provided in the Answer section for scoring the performance task portion of each test.

- **Ideas for Use** Each benchmark test can be used as a diagnostic assessment prior to instruction or as a summative assessment upon completion of instruction. Forms A and B can be used as a pretest at the beginning of the year and then as a posttest at the end of the year to measure mastery progress.

Go Online for More! connectED.mcgraw-hill.com

Performance Task rubrics to help students guide their responses are also available. These describe the tasks students should perform correctly in order to receive maximum credit.

Additional year-end performance tasks are available for Grades K through 5 as blackline masters, available under Assessment in ConnectED.

Students can also be assigned tech-enhanced questions from the eAssessment Suite in ConnectED. These questions provide not only rigor, but the functionality students may experience when taking the online assessment.

Assessment Item Types

In the spring, you will probably take a state test for math that is taken on a computer. The problems on the next few pages show you the kinds of questions you might have to answer and what to do to show your answer on the computer.

Selected Response means that you are given answers from which you can choose.

Selected Response Items

Regular multiple choice questions are like tests you may have taken before. Read the question and then choose the <u>one</u> best answer.

> **Multiple Choice**

A teacher gives 8 students some marbles to play a game. She has 70 marbles total. The teacher gives each student 1 marble until all 70 marbles are gone. How many students get exactly 9 marbles?

- ☐ 2
- ☐ 3
- ☐ 4
- ☐ 6

ONLINE EXPERIENCE Click on the box to select the one correct answer.

HELPFUL HINT Only one answer is correct. You may be able to rule out some of the answer choices because they are unreasonable.

> **Try On Your Own!**

Amy uses $\frac{3}{4}$ of a yard of fabric for each pillowcase that she makes. How many yards of fabric will she need in order to make 9 pillow cases?

- ☐ $5\frac{1}{4}$
- ☐ $6\frac{3}{4}$
- ☐ $7\frac{3}{4}$
- ☐ $8\frac{1}{4}$

Assessment Item Types

Sometimes a multiple choice question may have more than one answer that is correct. The question may or may not tell you how many to choose.

> **Multiple Correct Answers**

Select **all** the numbers that make this inequality true.

$$2\frac{1}{5} > \frac{1}{5} + \square + 1$$

- ☐ $\frac{1}{5}$
- ☐ $\frac{3}{5}$
- ☐ $\frac{5}{5}$
- ☐ $\frac{10}{5}$
- ☐ $\frac{4}{5}$

ONLINE EXPERIENCE Click on the box to select it.

HELPFUL HINT Read each answer choice carefully. There may be more than one right answer.

> **Try On Your Own!**

Select **all** equations that are true.

- ☐ $\frac{4}{10} = 0.4$
- ☐ $\frac{24}{100} = 2.4$
- ☐ $\frac{6}{100} = 0.6$
- ☐ $\frac{5}{100} = 0.05$

Assessment Item Types

Another type of question asks you to tell whether the sentence given is true or false. It may also ask you whether you agree with the statement, or if it is true. Then you select yes or no to tell whether you agree.

Multiple True/False or Multiple Yes/No

$\frac{4}{6}$ of the rectangle is shaded gray.

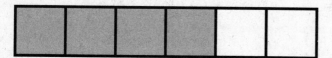

Decide if each fraction is equal to $\frac{4}{6}$. Select Yes or No for each fraction.

⏻ ONLINE EXPERIENCE
Click on the box to select it.

💡 HELPFUL HINT There is more than one statement. Any or all of them may be correct.

Yes	No	
☐	☐	$\frac{6}{9}$
☐	☐	$\frac{7}{12}$
☐	☐	$\frac{1}{2}$
☐	☐	$\frac{2}{3}$

▶ Try On Your Own!

Select True or False for each comparison

True	False	
☐	☐	$\frac{5}{6} > \frac{3}{5}$
☐	☐	$\frac{4}{10} > \frac{5}{8}$
☐	☐	$\frac{4}{6} < \frac{3}{4}$
☐	☐	$\frac{2}{3} > \frac{7}{9}$

Assessment Item Types **ix**

You may have to choose your answer from a group of objects.

Click to Select

Brayden and Ann both collect coins. Brayden has 86 coins. Ann has four times as many coins as Brayden. How many coins does Ann have? Select all the equations that represent this problem.

86 ÷ 4 = ☐   4 × 86 = ☐

4 × ☐ = 86   ☐ ÷ 4 = 86

☐ ÷ 86 = 4   86 ÷ ☐ = 4

⏻ **ONLINE EXPERIENCE**
Click on the item to select it.

💡 **HELPFUL HINT**
On this page you can draw a circle or a box around the item you want to choose.

▶ **Try On Your Own!**

Select **all** figures that have line symmetry.

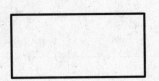

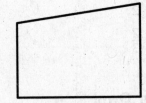
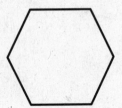

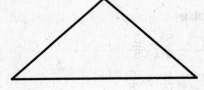

X   Assessment Item Types

When no choices are given from which you can choose, you must create the correct answer. These are called **constructed-response** questions.

One way to answer is to type in the correct answer.

Constructed-Response Items

Fill in the Blank

A pattern is generated using this rule: Start with the number 46 as the first term and subtract 7 from each term to find the next. Enter numbers into the boxes to complete the table.

Term	Number
First	46
Second	
Third	
Fourth	
Fifth	

> **ONLINE EXPERIENCE**
> You will click on the space and a keyboard will appear for you to use to write the numbers and symbols you need.
>
> **HELPFUL HINT**
> Be sure to provide an answer for each space in the table.

▶ **Try On Your Own!**

Jorge and his two brothers each water the plants. Jorge's watering can holds half as much water as his older brother's watering can. His younger brother's watering can holds 6 cups. Altogether the watering cans hold 3 gallons. How many cups does Jorge's watering can hold?

Assessment Item Types **xi**

Sometimes you must use your mouse to click on an object and drag it to the correct place to create your answer.

Drag and Drop

Drag one fraction to each box to make the statements true.

 =

ONLINE EXPERIENCE
You will click on a number and drag it to the spot it belongs.

HELPFUL HINT
Either draw a line to show where the number goes or write the number in the blank.

 =

$2\frac{1}{3}$ $4\frac{1}{3}$ $3\frac{2}{3}$ $\frac{10}{3}$ $\frac{7}{3}$ $\frac{13}{3}$

Try On Your Own!

Order from least to greatest by dragging each fraction to a box.

$\frac{4}{6}$ $\frac{1}{3}$ $\frac{7}{8}$ $\frac{4}{4}$ $\frac{3}{4}$ $\frac{1}{2}$

Some questions have two or more parts to answer. Each part might be a different type of question.

Multipart Question

Each bag of dried apples has 4 servings. Each bag of dried bananas has 5 servings. How many servings of dried fruit are in 7 bags of dried apples and 3 bags of dried bananas?

Part A: Drag the numbers to the boxes and the symbols to the circles to make an equation to show how many servings of dried fruit there are in all.

Part B: Select the correct number of servings of dried fruit Anna has.

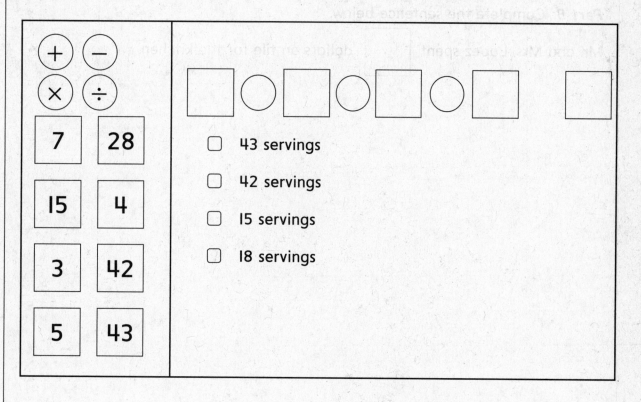

- 43 servings
- 42 servings
- 15 servings
- 18 servings

Assessment Item Types xiii

▶ **Try On Your Own!**

Mr. and Mrs. Lopez are putting tiles on the floor in their kitchen. They can fit 12 rows of 14 tiles in the kitchen. If each tile costs $3, what is the total cost?

Part A: Choose a sentence to find how many tiles they needed in all.

- ☐ 12 + 14 = 26 tiles
- ☐ 12 × 2 + 14 × 2 = 52 tiles
- ☐ 12 × 14 = 168 tiles
- ☐ 12 × 2 + 14 = 38 tiles

Part B: Complete the sentence below.

Mr. and Mrs. Lopez spent ☐ dollars on tile for the kitchen.

NAME ....

DATE ....

SCORE ....

Countdown: 20 Weeks

1. Use the clues to complete the place-value chart.

Thousands			Ones		
hundreds	tens	ones	hundreds	tens	ones

> **ONLINE TESTING**
> On the actual test, you might be asked to drag the numbers into the correct places on the place-value chart. In this book, you will be asked to write the numbers in the chart..

The 2 has a value of 2 × 10.

The 3 has a value of 3 × 100,000.

The 4 is in the ones place.

The 5 is in the hundreds place.

The 6 is in the ten thousands place.

The 8 has a value of 8 × 1,000.

2. Circle the statement that shows *one hundred seventy-five thousand, two hundred sixty four* in expanded form.

   A. 100,000 + 70,000 + 5,000 + 200 + 60 + 4

   B. 175,264

   C. 264,175

   D. 200,000 + 60,000 + 4,000 + 100 + 70 + 5

3. Compare. Use >, <, or =.

   Part A: 689,674 ___ 689,476

   Part B: 264,864 ___ 264,864

4. Circle the answer that shows 546,216 rounded to the nearest ten thousands place.

   A. 500,000

   B. 546,200

   C. 556,216

   D. 550,000

5. Order the numbers from greatest to least.

   354,678

   354,687

   345,876

Countdown: 19 Weeks

1. **Part A:** Find the unknown.

    57 + (8 + ____) = (57 + 8) + 9

    Part B: Circle the addition property it shows.

    A. Associative Property

    B. Commutative Property

    C. Identity Property

    D. Parentheses Property

2. Draw a line connecting the number to its description.

1,000 less than 85,678	85,875
200 more than 85,675	84,678
10,000 less than 85,675	75,675
10,000 more than 85,675	95,675

    ONLINE TESTING
    On the actual test, you might be asked to drag and drop the numbers. In this book, you will be asked to write the answers.

3. Circle the answers that show ways of estimating the addition problem below.

   46,417
   + 71,654
   ―――――

   A. 72,000
     + 46,000
       ―――――

   B. 70,000
     + 40,000
       ―――――

   C. 71,654
     + 46,417
       ―――――

   D. 71,700
     + 46,400
       ―――――

4. Subtract. Use addition to check.

   798,656
   − 465,684
   ―――――――

5. Write the number that completes the number sentence.

   51,268 − _____ = 41,268

NAME _____   DATE _____

Countdown: 18 Weeks

SCORE _____

1. The table shows the production numbers for a company that manufactures cell phones. The table entry is the number of phones manufactured in the given year.

Year	Number of Phones Manufactured
2010	2,345,000
2011	3,155,000
2012	4,050,000
2013	4,891,000

   Part A: Between which two years did the company experience the most growth in the number of cell phones they manufactured? What was the increase?

   Part B: Estimate to the nearest ten thousand the number of phones manufactured in the four-year period.

2. Mr. Harrison is saving money to pay for a family vacation in August. The total cost of the trip will be $1,195. In March, he saved $185. In April he saved $412. In May he save $204. In June he saved $175. Use the following numbers to estimate how much Mr. Harrison needs to save in July in order to pay for the vacation.

   ONLINE TESTING
   On the actual test, you might be asked to drag numbers to the correct location. In this book, you will be asked to write the numbers using a pencil.

   | 200 | 200 | 200 | 400 | 1,200 | 200 |

   ☐ − (☐ + ☐ + ☐ + ☐) = ☐

Grade 4 · Countdown 18 Weeks   **17**

3. Select **each** statement that is a member of the fact family for the array.

   ☐ 5 × 6 = 30

   ☐ 5 + 5 = 10

   ☐ 30 ÷ 6 = 5

   ☐ 30 ÷ 5 = 6

   ☐ 30 ÷ 10 = 3

4. Jamal earned $12 the first snowfall of the year for shoveling driveways. During the heaviest snowfall of the year, he earned 4 times the amount that he earned during the first snowfall. Use numbers and symbols from the list shown to write an equation that can be used to find a, the amount that Jamal earned during the heaviest snowfall of the year.

   3   4   8   12 a ×   ÷   +   −

   ┌───┐
   │   │
   └───┘

5. There are 12 students on a soccer team. Each student has 4 boxes of ornaments to sell for a fundraiser. Each box has 3 ornaments in it. Select **all** number statements that could be used to find out how many ornaments there are in all.

   ☐ (3 × 4) × 12

   ☐ 3 × (4 × 12)

   ☐ 4 × 12

   ☐ 3 × 12

   ☐ 12 × 12

   ☐ (3 + 4) × 12

18 Grade 4 · Countdown 18 Weeks

Countdown: 17 Weeks

1. A tray of muffins is arranged as shown. Draw one other way that the same muffins can be arranged into rows and columns

ONLINE TESTING
On the actual test, you may be asked to arrange the objects by clicking on them. In this book, you will instead draw the objects using a pencil.

2. A grocer has 32 boxes of rice cereal, 18 boxes of oat cereal, and 13 boxes of bran cereal. He wants to put these boxes into three different displays, and he wants each display to have the same number of boxes.

   Part A: Is it reasonable to state that there will be more than 15 boxes in each display? Explain.

   Part B: Write and solve an equation whose solution is *b*, the number of boxes in each display.

3. Two swimmers keep track of the number of laps they swim each day. For each day, tell whether Swimmer A swims 3 times as many laps as Swimmer B.

Day	Swimmer A	Swimmer B	Does A swim 3 times as many laps as B?	
			Yes	No
Monday	12	4		
Tuesday	3	9		
Wednesday	18	7		
Thursday	15	5		
Friday	9	6		

4. Quentin was solving a multiplication problem. Circle the names of **all** properties he used to solve the problem.

$$5 \times (6 \times 2) = 5 \times (2 \times 6)$$
$$= (5 \times 2) \times 6$$
$$= 10 \times 6$$
$$= 60$$

Associative Property of Multiplication   Commutative Property of Multiplication   Identity Property of Multiplication

5. The number 42,__ __2 rounded to the hundreds place is 42,300. What is the least possible sum of the two missing digits? Justify your answer.

20 Grade 4 • Countdown 17 Weeks

Countdown: 16 Weeks

1. Select **each** expression that represents the model shown.

   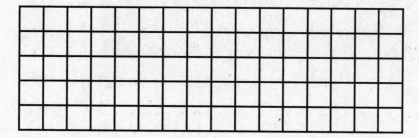

   ☐ (5 × 10) + (5 × 6)

   ☐ 5 × 16

   ☐ (5 × 10) + (6 × 10)

   ☐ (5 × 15) + 5

   > **ONLINE TESTING**
   > On the actual test, you might be asked to click on the boxes to select them. In this book, you will instead shade the boxes with a pencil.

2. The Suarez family is taking a vacation. They will drive 584 miles every day for three days.

   A. Estimate the number of miles the family will drive in total over the three days. Show your estimate.

   B. Will the exact number of miles be less than or greater than your estimate? Explain.

   C. Find the exact number of miles the family will drive over the three days.

Grade 4 · Countdown 16 Weeks **21**

3. Select Yes or No to indicate whether or not the product will require regrouping.

   Requires Regrouping?

   Yes No

   ☐   ☐   305
            × 9

   ☐   ☐   1,234
            × 2

   ☐   ☐   1,402
            × 3

4. Henry is looking at a set of base ten blocks.

   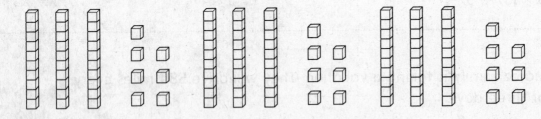

   A. What multiplication problem does this represent?

   [   ]

   B. What is the answer to the multiplication problem?

   [   ]

5. Use the Associative Property of Multiplication to solve the equation in the easiest way possible. Show your work.

   $5 \times (4 \times 14) = $ _____

   [   ]

Countdown: 15 Weeks

1. Mrs. Giovanni plants a small garden of tomatoes every year. She places the plants into 26 rows of 15 plants each.

   Part A: Which expression shows the total number of plants?

   A. $(26 \times 10) + (26 \times 5)$

   B. $(26 \times 10) + (26 \times 6)$

   C. $(15 \times 20) + (15 \times 5)$

   D. $(26 + 10) \times (26 + 5)$

   Part B: Find another arrangement that would result in the same number of plants.

2. Estimate each product by rounding the larger number to the greatest place value possible. Then place an X in the appropriate box to indicate whether the estimate is greater than or less than the actual product.

Product	Estimate	Product is ____ than the estimate	
		Less	Greater
205 × 8	1,600		
1,827 × 5	10,000		
3,127 × 4	12,000		
18,412 × 7	140,000		

   ONLINE TESTING
   On the actual test, you may be asked to click in the boxes to add the X's. In this book, you will instead write the X's using a pencil.

3. Manuel is making cookies for a bake sale. He makes 16 batches. Each batch has b cookies in it. The number b is an odd 2-digit number. Indicate whether each statement is true or false.

True	False	
☐	☐	The total number of cookies is odd.
☐	☐	The total number of cookies could be less than 100.
☐	☐	The number b could have a zero in the one's place.
☐	☐	The total number of cookies could be a multiple of 10.
☐	☐	The number equal to $16 \times b$ could be greater than 1,000.

4. Which number correctly completes the equation?

$$26 \times 72 = \underline{} + 40 + 420 + 12$$

A. 140

B. 1,400

C. 14,000

D. 140,000

5. Uma scored 8 goals in her first season of soccer and 24 goals in her second season. Write two different statements that relate the two numbers. The first statement should involve addition. The second statement should involve multiplication.

Addition Statement

Multiplication Statement

Countdown: 14 Weeks

1. A farmer has a garden that he is dividing into four different regions. The garden can be modeled using an area model as shown. All measurements are in feet.

   > **ONLINE TESTING**
   > On the actual test, you may be asked to click into the boxes in order to key in the answers. In this book, you will instead write the answers using a pencil.

   Part A: Fill in the area of each of the four regions.

   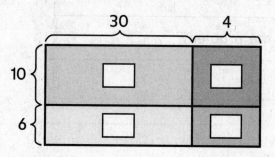

   Part B: What is the area of the entire garden?

2. Which of the following number sentences are not useful for finding 24 × 62?

   A. 20 × 2 = 40

   B. 24 × 2 = 48

   C. 24 × 6 = 144

   D. 4 × 2 = 8

3. Mrs. Shen is trying to write a multiplication problem for her fourth grade math class. She wants her students to expand two 2-digit numbers being multiplied together. She wants the answer to be:

$$(20 \times 50) + (20 \times 6) + (2 \times 50) + (2 \times 6)$$

Part A: What should the original multiplication problem be?

[ ]

Part B: What is the final answer to the problem?

[ ]

4. List three different ways of arranging 72 seats for a concert in rows and columns.

[ ]

5. A marathon runner runs 17 miles every day for training. He runs for d days and runs a total of 119 miles. Circle **all** of the following that express the relationship between 17, *d,* and 119.

$$17 \times d = 119$$

$$17 \times 119 = d$$

$$d \times 119 = 17$$

$$119 \div d = 17$$

$$119 \div 17 = d$$

$$17 \div d = 119$$

Countdown: 13 Weeks

1. An electronics company plans on producing 7,158 components in the next 12 days. The plan manager estimates that they should produce about 600 components per day. Select **each** true statement about the plant manager's calculation.

   > **ONLINE TESTING**
   > On the actual test, you may be asked to click on the boxes to select each answer. In this book, you will instead shade the boxes using a pencil.

   ☐ The manager rounded to the nearest hundred before dividing.

   ☐ The manager used compatible numbers to estimate.

   ☐ The manager's estimation can be checked with the product 600 × 12.

   ☐ The manager's estimation can be checked by adding 28 to 7,100.

   ☐ The manager's estimate is less than the actual number of components produced on a given day if they produce the same number of components each day.

2. A baker baked 6 types of bread. He baked the same number of each type. There are 360 loaves of bread total. How many loaves of each type did the baker bake?

   [   ]

3. A house painter has a goal of painting 500 rooms in 8 months. He uses the long division 8)500 to determine how many rooms he needs to paint each month. Determine whether each statement below is true or false.

   True False

   ☐   ☐   His problem has a remainder.

   ☐   ☐   If he paints 63 rooms each month, he will meet his goal.

   ☐   ☐   If he paints 62 rooms each month, he will meet his goal.

4. Four friends raised a total of $732 for a local charity. Each friend raised the same amount of money.

   Part A: Model the total amount that the friends raised using the least number of base-ten blocks possible.

   Part B: Show how to use the base-ten blocks to determine how much each friend raised. You may trade for different base-ten blocks, if necessary.

---

5. An electrician orders 48 packages of circuit breakers. Each packages contains 17 breakers.

   Part A: Write the correct numbers and symbols in the boxes to make an equation that shows an estimate of the number of circuit breakers ordered. Select from the list below.

   +   −   ÷   ×

   20   30   40   50   200   300   400   500   800   1,000

   ☐ ☐ ☐ = ☐

   Part B: Is the estimated number of circuit breakers ordered greater than or less than the actual number? Explain how you know.

Countdown: 12 Weeks

1. A woodworker is building 5 tables. He wants to figure out how many feet of wood each table can have. He has 603 feet of board. The woodworker will check his work. Write the numbers 1-3 next to the correct three steps below to show the order in which the woodworker should perform them.

   > **ONLINE TESTING**
   > On the actual test, you may be asked to drag the steps into order. In this book, you will instead number the steps using a pencil.

   _____ Add 2 to 601.

   _____ Add 3 to 120.

   _____ Divide 120 by 5.

   _____ Divide 603 by 5.

   _____ Multiply 121 by 5.

   _____ Multiply 120 by 5.

   _____ Add 3 to 600.

   _____ Subtract 3 from 600.

2. A florist orders 820 roses for a reception. Four of the roses were too damaged to use. The rest of the roses will go into vases holding 8 roses each.

   Part A: Write an equation that can be used to find *v*, the number of vases that the florist will need.

   Part B: Think about the numbers in the division problem. Explain how you can solve the problem using place value and mental math. Then state the answer.

3. A grocer is ordering juice containers that come in packages that have 7 bottles. The grocer starts writing out multiples of 7 in order to place a juice order. Select all correct statements.

- ☐ All of the numbers on the list are odd.
- ☐ The first ten multiples will contain every possible digit in the one's place before they start repeating.
- ☐ The sum of the digits in each number in the list is divisible by 7.
- ☐ Each number in the list is divisible by 7.
- ☐ Each number in the list is divisible by 14.

4. Xavier practices 10 free throws on Monday, 20 on Tuesday, 40 on Wednesday, and 80 on Thursday.

Part A: Describe the pattern.

Part B: Describe why this pattern is not reasonable for Xavier to continue by looking at the next three days.

5. The first term in a number pattern is 20. Additional terms are obtained by multiplying the previous term by 4. What is the first term that is larger than 1,000? Enter numbers into the table to figure this out. Stop when you get a term greater than 1,000.

Term	Number
First	20
Second	
Third	
Fourth	
Fifth	

NAME ....

DATE ....

SCORE ....

Countdown: 11 Weeks

1. In the table shown, each rule's input is the previous rule's output.

   Part A: Complete the table by writing in the missing values.

Input	Add 3	Subtract 5	Multiply by 5
5	8	3	
6			
	18		
			30

   Part B: If n is the input, write an equation that can be used to find p, the output.

2. A scientist is studying an ant infestation at a local building site. On the first day, he observes 16 ants along the base of the entryway. On the second day, he observes 20 ants. On the third, he observes 24 ants. This pattern continues for the rest of the week. Indicate whether **each** statement is true or false.

   ONLINE TESTING
   On the actual test, you may be asked to click into the boxes to shade them. In this book, you will instead shade the boxes using a pencil.

   True False
   ☐ ☐ The number of ants is always a multiple of 4.
   ☐ ☐ The rule for the number of ants seen the next day is "add 8".
   ☐ ☐ The number of ants is always a multiple of 8.
   ☐ ☐ The number of ants is always a multiple of 2.
   ☐ ☐ The total number of ants for the week is a multiple of 4.

3. How many branches will be at the bottom of the 7th figure in the pattern shown?

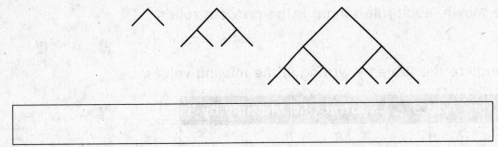

4. A large pot of water contains 64 cups of water. As the water boils, the amount of water in the pot decreases. After 10 minutes, there are 48 cups left. After another 10 minutes there are only 32 cups left. If this pattern continues, circle the number of minutes it will take until all of the water is gone.

   A. 4 minutes   C. 30 minutes

   B. 40 minutes   D. 50 minutes

5. A store is putting together packages of school supplies to donate to a local school. The table shows the number of each type of each supply they have to donate.

Type of Supply	Number of Supply
Pencils	172
Pens	161
Packs of Paper	45
Book Bags	31

Part A: How many boxes can be packed if each package must contain 10 writing utensils (pens or pencils)?

Part B: If each package will be put into a book bag, are there enough book bags to fit your answer from *Part A*? Explain.

32 Grade 4 · Countdown 11 Weeks

Countdown: 10 Weeks

1. Dr. Nicole Soto has developed Ivy-X, a poison ivy remedy that she thinks works better than the standard treatment. Dr. Soto's data is shown.

	Got better	Got worse
Ivy-X treatment	12	4
Standard treatment	10	5

   ONLINE TESTING
   Sometimes you can take values directly from a table as the solution to your problem. In other cases, you may need to take the values from the table and modify them before you can find your solution.

   Part A: What fraction of the patients who used Ivy-X got better? What fraction got worse?

   Part B: What fraction of the patients who used the standard treatment got better? What fraction got worse?

   Part C: Is Dr. Soto correct? Does Ivy-X actually work better than the standard poison ivy treatment? Explain.

2. Ricky planned to spend $20 to enter Sky Rock Water Park and buy extra tickets for 4 special rides while he is there. The special rides cost $3 each. Ricky wrote this expression to show the total amount of money he needs to bring.

   $$\$20 + 4 \times \$3$$

   Using the expression above, Ricky calculates that he needs to bring a total of $72 to the park. Is Ricky correct? Explain. If he is not correct, how much does he need to bring?

3. Craig told Dava the first two numbers in his locker combination. To find the third number Craig said, "Find the sum of all of the factors of the number 20."

   What are the factors of the number 20? What is the third number of Craig's combination?

4. Explain why the model does or does not show $\frac{3}{8}$.

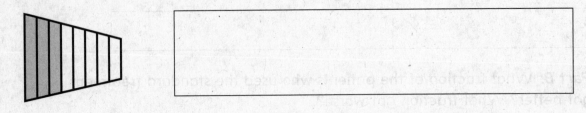

5. What mixed number is equivalent to $\frac{24}{5}$? How can you use models to show that your answer is correct? Explain.

Countdown: 9 Weeks

1. The attendance at a town hall meeting is expected to be 75. Folding chairs for the meeting are arranged in rows of 6.

    Part A: How many rows are needed to fit all of the people?

    Part B: If exactly 75 people show up for the meeting, how many seats will be left empty?

2. Between 0 and 99, the greatest number of prime numbers end in which digit — 0, 1, 2, 3, 4, 5, 6, 7, 8, or 9? Explain.

    ONLINE TESTING
    For problems like this one you might find it helpful to use the problem-solving strategy of make a list. Use this strategy whenever you have a group of items from which you need to find a pattern.

3. Kalief claims that he can carry out any mathematical operation forward or backward. As an example, Kalief shows the equations below. Is Kalief correct? Explain.

    $7 + 4 = 11$
    $4 + 7 = 11$

4. In basketball, a player is considered an "excellent" shooter if he or she makes more than half of his or her shots.

Player	Shots made	Shots taken
Steve	10	16
Boris	12	24
Kim	15	35
Shaniqua	25	40

Part A: Which of these players would be considered an excellent shooter? Explain.

Part B: Who was the best shooter in the group? Explain.

5. Explain why each model does or does not show $\frac{8}{3}$.

A.

B.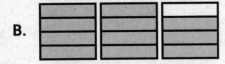

C.

D.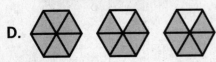

Countdown: 8 Weeks

1. Every day Rafael runs around the 1-mile track that is divided into 8 sections of equal length.

   Part A: From the starting line Rafael ran $\frac{11}{8}$ of a mile in the direction shown. Circle the line that shows the location on the track where Rafael ended his run.

   > **ONLINE TESTING**
   > On the actual test, you may need to click on the screen to show a location on the diagram. In this book, you will use your pencil to circle the location.

   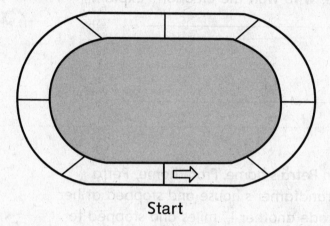

   Start

   Part B: From where Rafael stopped, how much farther does he need to run to reach the starting line again?

2. Bonnie put 5 gold coins that each weighed $\frac{2}{7}$ of a pound into a cloth sack. To the sack Bonnie added 4 silver coins that weighed $\frac{3}{7}$ of a pound each. What was the total weight of the coins in the sack? Explain how you found this total.

3. In the election for mayor, Victor Peña got 142,512 votes and Marley Robins received 142,483 votes.

   Part A: Who won the election and by how much?

   Part B: The election rules state that all results must be rounded to the nearest hundred. According the rule, who won the election? Explain.

4. Petra's grandfather lives 6 miles from Petra's home. From home, Petra rode her bike $3\frac{2}{5}$ miles toward her grandfather's house and stopped at her friend Hasan's. From Hasan's, Petra rode another $1\frac{4}{5}$ miles and stopped to see her friend Alfie. At Alfie's, how far from is Petra from her grandfather's house? Use the diagram to help explain how you got your answer.

   | $3\frac{2}{5}$ | $1\frac{4}{5}$ | |

   6

5. Which of following are equivalent to $4\frac{3}{8}$? Select all that apply.

   ☐ $7 \times \frac{5}{8}$

   ☐ $6 - 2\frac{5}{8}$

   ☐ $3\frac{5}{8} + 1\frac{6}{8}$

   ☐ $35 \times \frac{1}{8}$

Countdown: 7 Weeks

1. A radio station runs three different types of commercials each hour. The commercials each last a different amount of time.

Commercial type (minutes)	Short ($\frac{1}{6}$ min)	Medium ($\frac{1}{4}$ min)	Long ($\frac{2}{3}$ min)
Commercials per hour	10	12	8
Total time (min)			

Part A: How many minutes does each commercial type take up per hour on the station? Complete the table.

Part B: The station manager claims that the station runs less than 10 minutes of commercials each hour. Is she correct? Explain.

2. In the right column, write the values from the box from that are equivalent to each expression in the left column of the table.

$4 \times \frac{4}{9}$	
$\frac{12}{5}$	
$5 \times \frac{2}{3}$	

Box

$10 \times \frac{1}{3}$, $4 \times \frac{3}{5}$, $12 \times \frac{1}{5}$, $\frac{16}{9}$, $3\frac{1}{3}$, $8 \times \frac{2}{9}$

ONLINE TESTING
On the actual test, you may drag items to solve a problem. In this book, you will use your pencil to write in the items.

3. Allie picked 8 pounds of strawberries, which was far more than she or her family could eat. Allie gave $1\frac{3}{10}$ pounds to Dov, who said he was going to make a pie. Allie gave $2\frac{9}{10}$ pounds to Nina, who was making strawberry shortcake. After giving away strawberries to Dov and Nina, how many pounds of strawberries did Allie have left? Explain.

4. Tyler collected 8 butterflies. Flor collected 3 times as many butterflies as Tyler. Misha collected 5 times as many butterflies as Flor. How many more butterflies did Misha collect than Tyler?

5. What product does this model represent? Fill in the labels for the model and explain how the model helps you find the product.

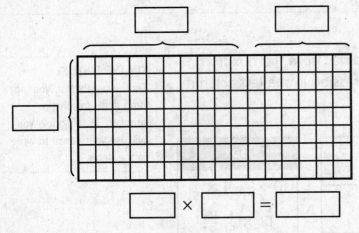

NAME

DATE

SCORE

Countdown: 6 Weeks

1. The five judges at the skateboard competition each gave out the scores to Tracee as shown. Write each score into the table from low to high.

   Judge 1: $\frac{9}{10}$ Judge 4: $\frac{27}{100}$

   Judge 2: $\frac{9}{100}$ Judge 5: 0.08

   Judge 3: 0.42

   > **ONLINE TESTING**
   > On the actual test, you may drag items to show your answer. In this book, you will use your pencil to write in the items.

   To get a final score, the high and the low scores are thrown out. The middle three scores are then added to give the final score. What was Tracee's final score? Write your answer as a decimal.

5 (low)	4	3	2	1 (high)

2. Which models correctly show the sum of $\frac{30}{100}$ + 0.2? Select all that apply.

    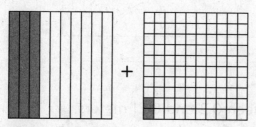   

    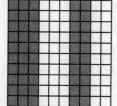   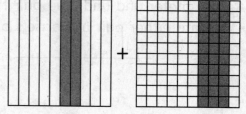

3. Results from the Catch and Release Bass Fishing Contest are shown.

Name	Doug	Kendal	Monk	Daphne
Length (m)	0.54	$\frac{6}{10}$	$\frac{9}{100}$	0.5

Who was the winner of the contest? Explain how you know.

4. Renting a canoe costs $6 per hour. All renters must also pay $3 for the paddle and life jacket. Zern wrote this equation for finding t, the cost of renting a canoe, for h hours.

$(6 \times h) + 3 = t$

Find the rental cost, t, when $h = 4$ hours. Show your work.

5. Bruce has measured his feet to be between 0.2 and 0.21 meters in length. Carrie has a pair of red bowling shoes that measure $\frac{23}{100}$ of a meter and and blue pair that are $\frac{18}{100}$ m in length. Can Bruce fit into either pair of Carrie's bowling shoes? Explain.

NAME	DATE
	SCORE

Countdown: 5 Weeks

1. Look at the following model.

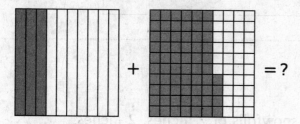

   Part A: Write a problem that would require you to use this model to find its solution.

   Part B: Solve the problem you wrote. Include a diagram and labels.

2. Ursula sold 47 copies her new app for $26 each. How much money did Ursula make? Use the numbers below to fill in the area model. You may not use all of the numbers.

   ONLINE TESTING
   On the actual test, you may drag items to show your answer. In this book, you will use your pencil to write in the items.

   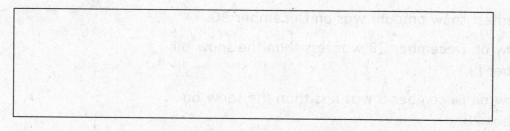

   800   36   42   6   20

   400   140   7   24   240

Grade 4 • Countdown 5 Weeks   43

3. Ruby has read 0.85 of a book. What fraction of the book does Ruby have left to read? Write your answer in simplest form.

4. It snowed five times in December with snowfalls of 7 inches, 2 inches, 6 inches, 11 inches, and 5 inches. Use the information below to fill in the table.

   - The smallest snow amount was on December 16.
   - The greatest snow amount was on December 30.
   - The snow on December 28 was less than the snow on December 14.
   - The snow on December 5 was less than the snow on December 28.

Date	Dec 5	Dec 14	Dec 16	Dec 28	Dec 30
Snow (inches)					

5. Which of the following is equivalent to $\frac{4}{10}$? Select all that apply.

   ☐ 0.25 + 0.15

   ☐ $\frac{2}{5}$

   ☐ 0.1 + 0.03

   ☐ $\frac{12}{100}$ + 0.28

44 Grade 4 · Countdown 5 Weeks

Countdown: 4 Weeks

1. Oscar figured out a way to multiply 34 × 18 using the Distributive Property and subtraction rather than addition. Oscar's idea is to multiply 34 by the "easy" number of 20 rather than 18. Then, to make up for adding 2 to 18 to make 20, Oscar will subtract 34 times 2 from the total. Fill in the blanks to show Oscar's method.

   34 × 18 = (_____ × 20) − (34 × _____)

   = (_____) − (_____)

   = _____

> **ONLINE TESTING**
> On the actual test, you may drag items to show your answer. In this book, you will use your pencil to write in the items.

2. Sundip conducted a survey to see how much time each classmate spends per week playing video games. Use Sundip's data to create a line plot. Make sure you include a scale for your line plot.

Less than 1 hr	1 to 2 hrs	2 to 3 hrs	3 to 4 hrs	4 to 5 hrs	More than 5 hrs
4	2	6	7	3	2

Grade 4 · Countdown 4 Weeks

3. Ned is building a wooden table. The table top has three sections and a total width of 8 feet. Ned has made the two outer pieces that measure 30 inches and 28 inches in width. How wide should the middle section be? Explain.

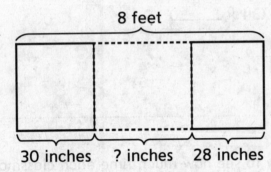

4. Which of the following is equivalent to 9 pints? Select all that apply.

   ☐ 172 fluid ounces   ☐ 1 gallon and 12 fluid ounces

   ☐ 3 quarts and 6 cups   ☐ 1 gallon and 1 pint

5. Without shoes, Morgan steps on a scale and weighs exactly 65 pounds. Morgan puts on a pair of sneakers and holds a basketball that weighs 23 ounces. Now she weighs $68\frac{1}{4}$ pounds. How much does each shoe weigh? Show your work.

NAME _____ DATE _____

Countdown: 3 Weeks

SCORE _____

1. The Z-Dog company is conducting a survey to help design its new tablet, the Z-Six. The Z-Six is designed for commuters who travel long distances to school or work each day. Circle the correct units for each question in Z-Dog's survey.

> **ONLINE TESTING**
> On the actual test, you will click on answers to multiple choice questions. In this book, you may be asked to circle answers to multiple choice questions.

   A. What approximate length should the Z-Six tablet have?

   millimeters   centimeters   meters   kilometers

   B. What exact width should the Z-Six tablet have?

   millimeters   centimeters   meters   kilometers

   C. How far do you travel to school or work?

   millimeters   centimeters   meters   kilometers

   D. What is the distance from your desk to the nearest electrical outlet?

   millimeters   centimeters   meters   kilometers

2. Jillian deposited a total of $120 in the bank over three days.

   Part A: On Tuesday Jillian deposited $\frac{2}{5}$ of the total. How much did she deposit on Tuesday?

   Part B: On Wednesday, Jillian deposited half of the total. How much did Jillian deposit on Wednesday?

   Part C: On Thursday, Jillian deposited the rest of the money. How much did Jillian deposit on Thursday? Explain.

Grade 4 • Countdown 3 Weeks   **47**

3. Nina wants to hang this picture so that the amount of space on each side of the picture is the same. How far from the end of the wall should Nina hang the picture? Show your work.

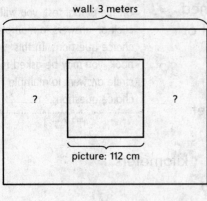

4. Which of the following is equivalent to 2.04 liters? Select all that apply.

- ☐ 204 milliliters
- ☐ 2 liters and 4 milliliters
- ☐ 2 liters and 40 milliliters
- ☐ 2,040 milliliters

5. The quotient of 350 ÷ 10 is equal to what number divided by 1,000?

NAME

DATE

SCORE

Countdown: 2 Weeks

1. Gina is building a garage that needs to have 96 square meters of space. The garage must have a width of at least 4 meters and be measured in whole meters.

   Part A: How many different plans can Gina use for the garage? List the dimensions of each plan.

   Part B: To have the smallest possible perimeter, which plan should Gina choose? Explain.

2. A rectangle has a length of 15 feet and a perimeter of 42 feet. Which of the following is true? Select all that apply.

   ☐ The figure is a square.

   ☐ The width of the rectangle is 12 feet.

   ☐ The area of the rectangle is 90 square feet.

   ☐ One side of the rectangle measures 6 feet.

   ONLINE TESTING
   On the actual test, you will click on answers to multiple choice questions. In this book, you will color in boxes to answer multiple choice questions.

Grade 4 • Countdown 2 Weeks **49**

3. Diego explained how he solved the problem below. First he subtracted 30 from 54 to get 24. Then he divided 24 by 3 to get 8 and then added 4 to 8 to get a final answer of 12. Is Diego's solution correct? Explain.

$$54 - 30 \div 3 + 4$$

4. If you continue the pattern of shaded squares surrounded by unshaded squares, at what point in the pattern does the area of the shaded squares become greater than the area of the unshaded squares? Explain.

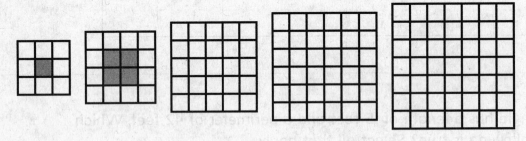

5. Milton started the morning with a full gallon of milk. He used 3 cups of milk to make biscuits and 3 pints of milk to make pudding. To make pancakes, Milton needs 6 cups of milk. Does he have enough milk to make the pancakes? Explain.

Countdown: 1 Week

1. The octagon below has 8 sides that are equal in length.

   Part A: How many lines of symmetry does the figure have? Draw them.

   [   ]

   Part B: How many equal-sized triangles do the lines of symmetry form?

   [   ]

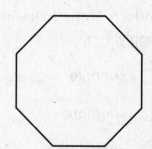

ONLINE TESTING
On the actual test, you will use your cursor to draw lines. In this book, you will draw lines using your pencil.

2. Match the numbers on the left to their factors on the right.

   40   12,6

   72   3,14

   56   8,5

   42   4,14

Grade 4 • Countdown 1 Week 51

3. What multiplication problem is represented by this diagram? Fill in the missing numbers. Write the product.

	☐	☐
☐	30 × 40	30 × 6
☐	4 × 40	4 × 6

4. Which of the following can have two obtuse angles? Select all that apply.

☐ triangle
☐ rhombus
☐ rectangle
☐ trapezoid
☐ parallelogram
☐ square

5. Angles a and b combine to form a right angle. The angle measure of a is 40° less than the angle measure of b.

What are the measures of the two angles? Use guess and check and the table to find the answer.

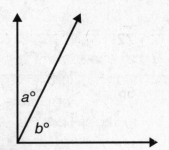

$a°$	$b°$	$a° + b°$	Is a 40° less than b?
10°	80°	90°	no, 70° less
40°	50°	90°	

$a =$ _____, $b =$ _____

52 Grade 4 • Countdown 1 Week

Chapter 1 Test

1. A scooter is priced between $1,000 and $2,000. Its price is a multiple of 10. All of the digits in the price, except for the thousands digit, are even numbers. The value of the hundreds digit is 30 times the value of the tens digit.

   Part A: Write the price of the scooter in standard form.

   Part B: Write the price of the scooter in word form.

2. The number on Ruben's ticket has a value between the values on the third and fourth raffle ticket shown.

   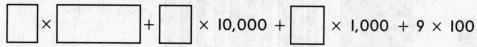

   739075 738901 739214 740000

   A partial expansion of Ruben's ticket number is shown. Fill in the blanks to complete the expansion.

   ☐ × ☐ + ☐ × 10,000 + ☐ × 1,000 + 9 × 100

3. In one month, a store sold between $3,000,000 and $4,000,000 worth of groceries. The estimate for the total sales that month had a 3 in the millions place, a 2 in the hundred thousands place, a 6 in the thousands place, and zeros in all of the other places.

   Part A: To what place was the total sales number most likely rounded? Explain your reasoning.

   Part B: What would be the estimate if the total sales were rounded to the nearest ten thousand?

4. The numbers of runners who participated in three long distance races are shown on the table. The number of runners in the Bolder Boulder Run was less the number in the Great North Run but more than the number in the New York City Marathon. Select all possible numbers of runners in the Bolder Boulder Run.

Race	Number of Runners
Peachtree Road Race	55,500
Great North Run	54,500
New York City Marathon	50,740

☐ 54,040   ☐ 55,000

☐ 50,697   ☐ 51,681

☐ 54,600   ☐ 50,089

5. Use some of the base-ten blocks to model 1,323 in two ways.

ONLINE TESTING
On the actual test, you might be asked to drag objects to their correct location. In this book, you will be asked to write or draw the object.

54 Grade 4 • Chapter 1 Place Value

6. Look at the table.

   Part A: Order the states from greatest (1) to least (4) according to number of acres of wilderness land area.

State	Wilderness Land Area (Acres)	Order
California	14,944,697	
Colorado	3,699,235	
Idaho	4,522,506	
Minnesota	819,121	

   Part B: Explain how you used place value to order the states.

7. A number is between 1 million and 8 million. The digit in the thousands place of the number is moved to a different position in the number so that the number's total value is 10 times greater.

   Part A: To which position did the digit move?

   Part B: After the digit is moved, the number is rounded to the nearest million. The result is greater than the original number. List all possibilities for the digit in the hundred thousands place of the original number. Explain your response.

Grade 4 · Chapter 1 Place Value 55

8. More than eight hundred fifty-four thousand people live in a city. Select all numbers that could be the population of the city.

☐ 8 × 100,000 + 55 × 1,000

☐ 8 × 100,000 + 6 × 10,000 + 1 × 1,000

☐ 807,230

☐ 98,727

☐ eight thousand, five hundred sixty-three

☐ nine hundred thousand

9. The table shows the capacities of four football stadiums.

Professional Football Team Stadiums	
Team	Stadium Capacity
Atlanta Falcons	71,228
Baltimore Ravens	71,008
Houston Texans	71,054
San Diego Chargers	70,561

Part A: Write the team names from least to greatest according to stadium capacity.

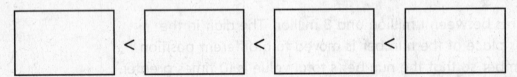

Part B: Which team's capacity has a digit with a value 10 times the value of the tens place digit in the Houston Texan's stadium capacity?

10. Three students round the same number to different places. Hector rounds to the thousands place and gets 15,000. Senthil rounds to the hundreds place and gets 14,700. Seiko rounds to the tens place and gets 14,650. What is the largest number that can be rounded as described?

11. The numbers of several types of apples harvested on a farm are shown. Select **Right** or **Left** to describe the placement of the indicated number on a number line.

Type of Apple	Number Sold
Gala	316,489
Fuji	309,500
Macintosh	312,740
Honeycrisp	297,081

Left Right

☐ ☐ The number of Macintosh apples is to the _____ of the number of Fuji.

☐ ☐ The number of Honeycrisp apples is to the _____ of 300,000.

☐ ☐ The number of Gala apples is to the _____ of the number of Macintosh.

☐ ☐ The number of Macintosh apples is to the _____ of 300,000.

☐ ☐ The number of Fuji apples is to the _____ of the number of Gala.

☐ ☐ The number of Gala apples is to the _____ of 320,000.

12. Jessica is playing a game. The goal of the game is to use 7 number cubes to create the greatest possible number. Jessica rolls the numbers shown.

Part A: What is the greatest number that Jessica could create by rearranging the number cubes?

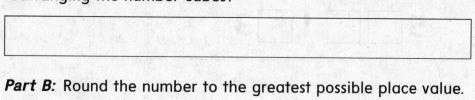

Part B: Round the number to the greatest possible place value.

13. Mrs. Norman's class recycled 3,412 pounds of newspaper. Mr. Rivera's class recycled 4,821 pounds of newspaper. Identify each digit whose value is 10 times greater in Mr. Rivera's number than in Mrs. Norman's number.

14. Jerome's city has a population of 1,567,231. Nhu's city has a population of 621,317. Select whether each statement is true or false.

    True False

    ☐ ☐ The 6 in Nhu's city population has 10 times the value of the same digit in Jerome's.

    ☐ ☐ The leftmost 1 in Jerome's city population has 10 times the value of the leftmost 1 in Nhu's.

    ☐ ☐ The 2 in Nhu's city population has 10 times the value of the same digit in Jerome's.

    ☐ ☐ The 3 in Jerome's city population has 10 times the value of the same digit in Nhu's.

    ☐ ☐ The cities' populations would have the same value in the hundred thousands place if they were rounded to the nearest hundred thousand.

15. In the number shown, the digit in the ten thousands place is 4 times the digit in the tens place. The digit in the tens place is an even number and is not equal to 0. The value of the digit in the thousands place has 10 times the value of the digit in the hundreds place. What is the complete number written in word form?

    5☐☐,1☐3

58 Grade 4 · Chapter 1 Place Value

Chapter 2 Test

1. Use the digits 1, 2, 3, 4, 5, and 6 exactly once to create a subtraction problem between two 3-digit numbers with the greatest possible difference. Then round the difference to the nearest hundred.

   [   ]

2. The table shows the number of laptops sold by Computer World each month during the second half of the year. Select **all** pairs of months that have sales numbers which add to an estimate of 60,000 laptops.

   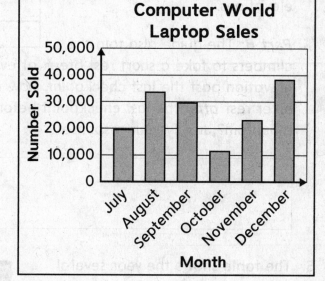
   Computer World Laptop Sales

   ☐ July and August

   ☐ July and December

   ☐ August and September

   ☐ August and November

   ☐ October and December

   ☐ November and December

3. David plans to run for 150 minutes every week. He ran for 28 minutes on Monday, 43 minutes on Tuesday, 39 minutes on Thursday, and 18 minutes on Friday. He estimates that he has about 20 minutes more to run to meet his goal. Write the correct numbers in the boxes to make an equation that shows how David could have come up with his estimate.

   ONLINE TESTING
   On the actual test, you might be asked to drag objects to their correct location. In this book, you will be asked to write or draw the object.

   | 10 | 20 | 30 | 40 | 50 | 150 |

   ☐ − (☐ + ☐ + ☐ + ☐) = ☐

Grade 4 • Chapter 2 Add and Subtract Whole Numbers   **59**

4. Nina is climbing to the top of Mt. Kilimanjaro. The mountain is 19,341 feet tall. Her base camp is at 9,927 feet. The guide for Nina's group has set up checkpoints at every 1,000-foot gain in elevation after the base camp.

   Part A: Complete the table below to show the checkpoint elevations.

	Elevation (ft)
First Checkpoint	10,927
	11,927
	12,927
	13,927
	14,927
	15,927
	16,927
	17,927
Last Checkpoint	18,927

   Part B: The guide also told the climbers to take a short rest break at every 100-foot gain in elevation past the last checkpoint. How many times should each hiker rest *after* the last checkpoint before reaching the top of the mountain? Justify your response.

   Each hiker should rest 4 times. The last checkpoint is at 18,927 feet and the summit is 19,341 feet. 19,341 − 18,927 = 414 feet. Rest breaks occur at 19,027; 19,127; 19,227; and 19,327 feet — 4 rest breaks before reaching the top.

5. The table shows the year several states were admitted to the United States and the year that the states' first censuses were recorded. Sort the states by the time between the first census and the year admitted. Write the state names in the correct column of the table below.

State	Year Admitted to United States	First Census
Kentucky	1796	1790
Arkansas	1836	1810
Michigan	1837	1790
Texas	1845	1820
West Virginia	1863	1860
Colorado	1876	1860
Utah	1850	1896

Number of Years Between First Census and Year Admitted to United States			
Fewer than 10	Between 10 and 20	Between 20 and 30	More than 30
Kentucky, West Virginia	Colorado	Arkansas, Texas	Michigan, Utah

6. Three backpackers are sorting their shared gear so that weight of each person's backpack is within a certain range. Each backpack's maximum weight range and current weight is shown. There are four items remaining. Sort the items into the three backpacks so that their weights are within the given range.

Cooking Kit 7 pounds   Tent 10 pounds   Shovel 2 pounds   Lantern & Batteries 4 pounds

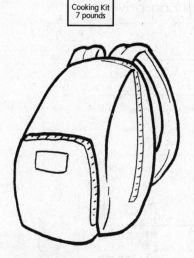
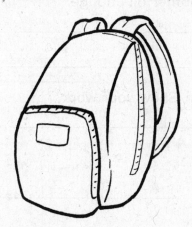
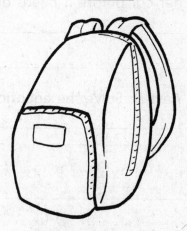

Backpack 1
Maximum Weight:
From 15 to 20 pounds
Current Weight:
12 pounds

Backpack 2
Maximum Weight:
From 20 to 25 pounds
Current Weight:
13 pounds

Backpack 3
Maximum Weight:
From 25 to 30 pounds
Current Weight:
25 pounds

7. A scientist recorded estimates of the number of bacteria in 4 colonies at the end of two 1-hour periods. Which colony had the greatest population growth, and by how many bacteria did it grow? Show your work.

Colony #	Hour 1	Hour 2
1	2,891,000	4,907,000
2	2,440,000	4,506,000
3	2,371,000	4,960,000
4	2,482,000	4,201,000

8. The number 38,☐☐2 rounded to the hundreds place is 38,700. What is the greatest possible sum of the two missing digits? Justify your answer.

Grade 4 · Chapter 2 Add and Subtract Whole Numbers

9. The owner's manual for Shanese's car recommends that she get her car's oil changed every 3,000 miles. Last month, Shanese had her car's oil changed and then drove 1,074 miles. She drove 883 miles this month.

   Part A: Write an equation with a letter standing for the unknown that can be used to find the number of miles Shanese can drive her car before it needs another oil change.

   Part B: Solve the equation. Show your work.

10. Ravi is saving money toward purchasing a bicycle that costs $350. He started this month with $78 in his savings account. He then saved $129 this month from his pay for delivering newspapers. His parents have agreed to pay $50 toward the cost of the bicycle. Select **all** expressions that are equal to the remaining amount that Ravi must save.

    ☐ 350 − 78 + 129 + 50 ☐ 350 − (78 + 129) + 50

    ☐ 350 − (78 − 129) − 50 ☐ 350 − (78 + 129 + 50)

    ☐ 350 − 78 − 129 − 50 ☐ 350 + 78 + 129 − 50

11. The table shows the current population of four cities and an estimate of the expected population increase or decrease over the next decade. Estimate each city's population, to the nearest thousand, at the end of the decade.

City	Population	Estimated Increase (+) or Decrease (−)
Belltown	31,897	+ 1,000
Claymore	121,043	− 5,000
Monroe	19,796	− 1,000
Stanton	63,662	+ 3,000

    Belltown: ☐   Monroe: ☐

    Claymore: ☐   Stanton: ☐

62 Grade 4 • Chapter 2 Add and Subtract Whole Numbers

12. The table shows the number of cars that can park in each of the four lots at a mall. The mall's owner plans use the parking lot that holds the least number of cars to build a parking garage that will hold from 1,000 to 1,200 cars.

Mall Parking Lots	
Parking Lot	Capacity
A	795
B	492
C	642
D	847

Part A: The garage will be built to hold the greatest number of cars possible. What will be the greatest number of cars that will be able to park at the mall? Justify your response.

Part B: The owner plans to build a second parking garage. What is the greatest number of cars that the second garage would need to hold so that a total of 4,500 cars can park at the mall? Justify your response.

13. Mr. Chavez took in $16,407 this month at his hardware store. He wrote a problem to subtract $9,154 he paid to his suppliers and $1,948 for expenses from what he took in. Shade in the box to select whether or not each statement describes a step that can be used to find how much money Mr. Chavez has left.

Yes	No	
		Use one hundred as 10 tens, which leaves 1 hundred and 15 tens.
		Subtract 4 tens from 4 tens to get 0 tens.
		Use one ten as 10 ones, which leaves 4 tens and 13 ones.
		Subtract 3 ones from 8 ones to get 5 ones.
		Use one thousand as 10 hundreds, which leaves 6 thousands and 12 ones.
		Subtract 9 hundreds from 12 hundreds to get 3 hundreds.

Grade 4 · Chapter 2 Add and Subtract Whole Numbers

14. A house decreased in value from $200,000 to $189,500. The value then decreased again by $2,900. What was the total decrease in the value of the house? Show your work.

15. The amount of tea left in four pitchers set out during a school meeting is shown at the right. The leftover tea was poured from all four pitchers into a 72-oz container. Shade the pitcher below to show the greatest amount of tea that can still fit into the 72-oz container.

16. The grid shows the seating chart for a school auditorium. The auditorium holds up to 225 people. The shaded cells show where adults are now seated to watch a play. In addition to the adults, there are currently 183 students in the auditorium. Select whether each statement is true (T) or false (F).

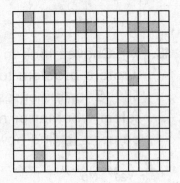

T	F	
		There are enough seats left for all of Mrs. Hill's 24 students.
		If all of Mr. Lee's 25 students were seated in the auditorium, all of the seats would be filled.
		There are enough seats left for all of Mr. Jackson's 28 students.
		If 8 more adults are seated, there will not be enough seats for all of Mr. Dearwater's 20 students.
		If 7 students leave their seats, there will be enough seats for all of Mrs. O'Dell's 32 students.

Chapter 3 Test

1. Beth has 15 seeds. She plants the same number of seeds in each of 3 flowerpots.

   Part A: Write a division statement whose answer is the number of seeds in each flowerpot. Include the answer in your statement.

   [   ]

   Part B: Represent the division statement on the number line.

2. Select **each** statement that is a member of the fact family for the array.

   - ☐ 42 − 6 = 36
   - ☐ 42 − 36 = 6
   - ☐ 42 ÷ 6 = 7
   - ☐ 42 ÷ 7 = 6
   - ☐ 6 × 6 = 36
   - ☐ 6 × 7 = 42
   - ☐ 7 × 6 = 42
   - ☐ 7 + 6 = 13
   - ☐ 6 + 7 = 13

3. A baker has 10 strawberries, 12 raspberries, and 10 blueberries. She plans to put the same total number of berries on each of 4 cakes.

   Part A: Before any calculations are done, is it reasonable to state that the baker will put fewer than 10 berries on each cake? Explain.

   Part B: Write and solve an equation whose solution is *b*, the number of berries on each cake.

4. Seats are arranged in a theater as shown. Each circle represents a seat. Show two more arrangements that use the same number of seats.

   ONLINE TESTING
   On the actual test, you might be asked to drag objects to their correct location. In this book, you will be asked to write or draw the objects.

5. Tanya divided cards from a game among her friends. Each friend received 13 cards. Tanya represented the division with the statement $52 \div \square = 13$.

   Part A: Write two multiplication facts for the division fact.

   [ ]

   Part B: How are the numbers in the multiplication facts related to the numbers in the division fact? Explain, using the appropriate vocabulary to identify each number.

   [ ]

6. Abby biked for 4 miles during the first week of summer vacation. During the last week, she biked 7 times as long as she did during the first week. Use numbers and symbols from the list shown to write an equation that can be used to find m, the number of miles that Abby biked during the last week of summer vacation.

   3   4   7   11 m ×   ÷   +   −

   [ ]

7. Dale divided 0 by a one-digit non-zero number and got a quotient of 0. Which number could have been the divisor? Explain.

   [ ]

8. Jay received 11 points on his first science project and 33 points on his second project. Write a statement that compares the number of points Jay received on the two projects. The solution to the statement should involve multiplication.

   [ ]

9. Two teams keep track of their points in a weekly math competition. For each week shown, indicate whether Team B has 4 times as many points as Team A.

Week	Team A	Team B	Does Team B have 4 times as many points as Team A?	
			Yes	No
1	||	HHT |||		
2			HHT		
3	HHT HHT ||	|||		
4	HHT ||||	HHT		
5	HHT HHT HHT								
6	HHT HHT	HHT			
7									
8	|||	HHT ||		

10. Marcus has 24 pencils, Melinda has 8 pencils, Jake has 5 pencils, and Suri has 32 pencils. For each pair of students, write a sentence comparing the number of pencils that they have.

Marcus and Melinda:

Melinda and Jake:

Suri and Marcus:

Suri and Melinda:

11. Each table in the science lab has 3 containers of bugs that are identical to the one shown. Each container has 5 bugs. Select **all** number statements that can be used to determine the total number of bug legs at each table.

☐ 6 × 3 × 5
☐ 5 × (3 × 5)
☐ (3 × 5) × 6
☐ 30 × 3
☐ 5 × 3
☐ 1 × 3 × 5

12. A bookcase has 4 shelves. There are 9 books on each shelf.

    Part A: Write two different multiplication sentences that show the total number of books in the bookcase.

    Part B: Write a third sentence that shows that the first two sentences are equal. Name the property that you used.

13. Eun earns 3 stickers for each news article that she reads. She earned 27 stickers for reading n articles. Select whether each statement is true or false.

True	False	
☐	☐	The equation 3 × 27 = n can be used to find the number of articles that Eun read.
☐	☐	The equation 3 × n = 27 can be used to find the number of articles that Eun read.
☐	☐	The equation 27 ÷ 3 = n can be used to find the number of articles that Eun read.
☐	☐	Eun read exactly 81 articles.
☐	☐	Eun read exactly 9 articles.
☐	☐	Eun read exactly 3 articles.

Grade 4 • Chapter 3 Understand Multiplication and Division

14. Sidney, Darius, Jorge, Karla, and Hasan each picked the same number of tomatoes from the school garden. The total number of tomatoes is shown. Circle groups of tomatoes to model the number that each student picked.

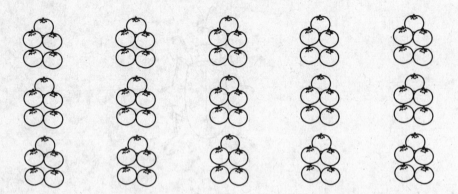

15. Kai found the product 8 × 3 × 1. First, he multiplied the 3 and the 1. Next, he multiplied that result by 8. Indicate whether each statement is true or false.

True	False	
☐	☐	Kai used the Commutative Property of Multiplication.
☐	☐	Kai used the Associative Property of Multiplication.
☐	☐	Kai used the Identity Property of Multiplication.
☐	☐	The result of the first operation that Kai performed is 24.
☐	☐	The final product is 24.

16. Miguel multiplied 0 by a one-digit non-zero number and got a product of 0. Which number could have been the other factor? Explain. Name the property that justifies this result.

70 Grade 4 · Chapter 3 Understand Multiplication and Division

Chapter 4 Test

1. Estimate each product by rounding the larger number to the greatest place value possible. Then check the appropriate box to indicate whether the estimate is greater than or less than the actual product.

Product	Estimate	Estimate is _____ than Actual Product	
		Less	Greater
119 × 7			
2,801 × 9			
169,032 × 4			
23,698 × 4			

2. Look at the base-ten blocks.

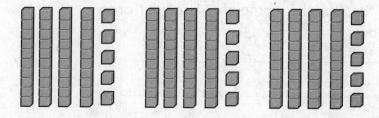

Part A: What product do the base-ten blocks represent?

Part B: Explain how to use the base-ten blocks and place value to find the product. Describe any regrouping that you must do.

3. The Ross family pays $105 each month for their cable and internet. Write the correct numbers in the expression to show the total amount, in dollars, that the Ross family pays for cable and internet in 6 months.

ONLINE TESTING
On the actual test, you might use a keyboard to type in answers. On this test, you will write the answers.

($\square \times \boxed{100}$) + ($\boxed{6} \times \square$) + ($\square \times \boxed{5}$)

4. As an assignment for art class, a group of art students completed information forms on different paintings. They visited 3 museums and completed 26 forms at each museum.

   Part A: Draw and label the area model to represent the total number of forms the students completed. Show the partial areas.

   Part B: How many forms did the students complete? Use your area model to explain how you know.

5. Four classes are going to see a play. Each class has 23 students.

   Part A: Use base-ten blocks to model the total number of student tickets needed by the four classes.

   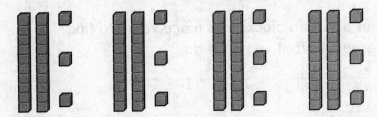

   Part B: How many tickets are needed? Explain how your model shows this.

6. Select **each** expression that represents the model shown.

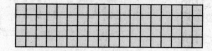

☐ (4 × 10) + (8 × 10)
☐ 4 × 10 × 8
☐ (4 × 10) + (4 × 8)
☐ 16 × 18
☐ (4 × 10) × (4 × 8)
☐ 4 × 18

7. Look at the area model.

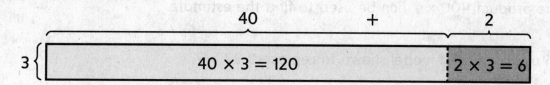

Part A: Write the partial products represented by the area model. Find the sum of the partial products.

Part B: Write the complete number sentence that is represented by the area model.

8. Lila estimated the product ☐,352 × 6 as 30,000 by first rounding to the greatest place value possible. What is the missing digit?

9. The tripmeter shows the number of miles that a delivery driver travels on one of his routes. He travels the route once each day on Monday, Tuesday, Wednesday, and Thursday. Select **each** statement that correctly describes the estimate of the number of miles the driver travels on the route each week.

☐ The estimate is less than the actual number of miles.

☐ The estimate is greater than the actual number of miles.

☐ The product 300 × 4 can be used to find the estimate.

☐ The product 300 × 7 can be used to find the estimate.

☐ The product 400 × 4 can be used to find the estimate.

☐ The product 400 × 7 can be used to find the estimate.

10. Dillon used the area model shown to represent 37 × 7.

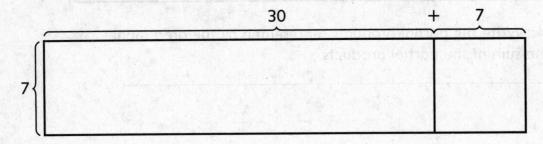

30 × 7 = 210

210 + 7 = 217

Part A: He used partial products as shown to calculate the product 37 × 7. What mistake did Dillon make?

Part B: Use partial products to find the actual product. Show your work.

74   Grade 4 • Chapter 4 Multiply with One-Digit Numbers

11. You want to multiply the numbers shown as they are written. For each product, select whether you would have to use regrouping.

Requires Regrouping?

Yes   No

☐   ☐   2,08
         × 5

☐   ☐   1,133
         × 2

☐   ☐   4,509
         × 3

> **ONLINE TESTING**
> On the actual test, you might click to select options. In this book, you will use your pencil to shade the correct boxes.

12. As part of a fundraiser, each grade level has a goal of raising $2,250. How much will the 6 grade levels raise if they all meet their goals? Show your work.

13. An assembly line in a factory fills 1,130 bottles of juice every hour. Indicate whether each statement is true or false.

True   False

☐   ☐   It is reasonable to state that the assembly line will meet its goal of filling 10,000 bottles during one 8-hour work day.

☐   ☐   A good estimate for the number of bottles filled during a 3-hour time period is 3,000.

☐   ☐   A supervisor would have to do some regrouping to find the number of bottles that the assembly line can fill in 5 hours.

☐   ☐   The assembly line can fill twice as many bottles in 4 hours as it can fill in 2 hours.

☐   ☐   The assembly line can fill at most 6,680 bottles during a 6-hour time period.

Grade 4 · Chapter 4 Multiply with One-Digit Numbers

14. Use your knowledge of place value and multiplication to find the missing digits in the product shown. Write the missing digits in the boxes.

$$1,6\boxed{}3 \times \boxed{} = 6,692$$

15. A bakery makes 2,094 loaves of bread each week. The manager calculated that the bakery produces 8,076 loaves in a 4-week period.

    Part A: What mistake did the manager make in calculating the total? Use place value in your explanation.

    Part B: What is the actual number of loaves that the bakery makes in a 4-week period? Show your work.

16. A local farm cooperative charges $98 per month for its family produce plan. Select **each** expression that can be used to find the amount a family must pay for a 6-month plan.

    ☐ (6 × 9) + (8 × 9)
    ☐ 98 × 6
    ☐ (6 × 90) + (6 × 8)
    ☐ (6 × 90) × (8 × 90)
    ☐ 6 × 8 + 6 × 90
    ☐ 90 × 8 + 6

NAME _____   DATE _____

Chapter 5 Test

SCORE _____

1. Mr. Gibson passed out the same number of cubes to each student in his class.

   - The number of cubes each student has is greater than 20.
   - Each student has an odd number of cubes.
   - Mr. Gibson passed out a total of 540 cubes.

   Select **all** statements that are true.

   ☐ Mr. Gibson could have 54 students in his class.

   ☐ Mr. Gibson could have 40 students in his class.

   ☐ Mr. Gibson has an even number of students in his class.

   ☐ The number of students in Mr. Gibson's class is less than 30.

   ☐ If Mr. Gibson had passed out 1080 total cubes, each student would have twice as many cubes.

> **ONLINE TESTING**
> On the actual test, you might click to select options. In this book, you will use a pencil to shade the correct boxes.

2. Estimate each product by rounding each number to the greatest place value possible. Then check the box to indicate whether the estimate is less than or greater than the actual product.

Product	Estimate	Estimate is _____ than Actual Product	
		Less	Greater
21 × 53			
89 × 64			
55 × 31			
72 × 40			

3. Which number sentence can you use as a first step to help you find 21 × 70?

   A. 21 + 7 = 28   C. 21 × 7 = 147

   B. 21 + 70 = 91   D. 21 ÷ 3 = 7

Grade 4 • Chapter 5 Multiply by a Two-Digit Number

4. A theater has 24 rows. Each row has 16 seats.

   Part A: Which expression shows the total number of seats in the theater?

   A. (20 × 10) + (4 × 6)   C. (20 × 6) + (4 × 10)

   B. (24 × 10) + (24 × 6)   D. (24 × 6) × (24 × 10)

   Part B: Complete the area model. Select from the numbers shown. Then write the total number of seats in the theater. The model should match the expression from **Part A**.

   4   6   10   16   20   24   40   200   144   240

   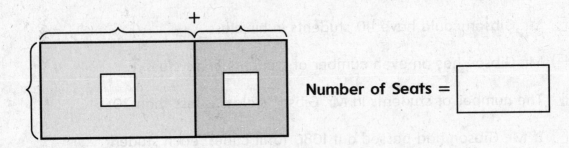

   Number of Seats = ▢

5. Ms. Gonzalez passed out some worksheets to her class. The total number that she passed out is shown by the area model.

   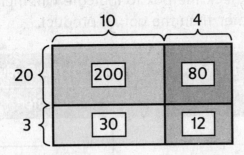

   Part A: Write an equation that shows the product that the area model represents.

   ▢

   Part B: Complete the statement below. Notice that there are two different ways that it could be completed.

   Ms. Gonzalez passed out ▢ worksheets to each of the ▢ students in her class.

   78 Grade 4 · Chapter 5 Multiply by a Two-Digit Number

6. Adele has 12 solid-colored marbles and 27 multicolored marbles in her collection. Wei has 13 times as many marbles in his collection. Select **all** equations that can be used to find m, the number of marbles in Wei's collection.

☐ $39 \div m = 13$

☐ $m = (12 \times 13) + (27 \times 13)$

☐ $m \div 13 = 39$

☐ $m = (12 \times 10) + (27 \times 3)$

☐ $m = 13 \times 39$

☐ $39 \div 13 = m$

7. A student made an error in the multiplication problem shown.

$$\begin{array}{r} 65 \\ \times\ 42 \\ \hline 130 \\ +260 \\ \hline 390 \end{array}$$

Part A: Explain the student's error, using place value.

Part B: Write a number in each of the boxes to correctly complete the problem.

8. Which number correctly completes the equation?

$34 \times 87 = 2400 + 210 + \boxed{} + 28$

A. 32

B. 320

C. 272

D. 2720

9. A parking lot has 15 rows. Each row has p parking spaces. The number p is an even 2-digit number. Indicate whether each statement is true or false.

   True False
   ☐   ☐   The lot has an even total number of parking spaces.
   ☐   ☐   The total number of parking spaces could be less than 100.
   ☐   ☐   The number p must have a zero in the one place.
   ☐   ☐   The total number of parking spaces is a multiple of 10.
   ☐   ☐   The number equal to 15 × p could be greater than 1,000.

10. The pictograph shows the number of cans collected by students at Riverview Elementary during a recycling drive. Write the correct number in each box to make an equation that models the situation.

    Number of Cans Collected

    | Grade 1 | 🥫🥫 |
    | Grade 2 | 🥫🥫🥫 |
    | Grade 3 | 🥫🥫🥫 |
    | Grade 4 | 🥫🥫🥫 |
    | Grade 5 | 🥫🥫🥫🥫 |
    | Grade 6 | 🥫🥫🥫 |
    | key: 🥫 = 25 Cans |

    ☐ × ☐ = ☐

11. Write the missing digits that complete the multiplication problem.

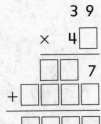

12. A recreation center orders 38 cartons of tennis balls. Each carton contains 6 of the containers shown in the picture.

    Part A: Write the correct numbers and symbols in the boxes to make an equation that shows an estimate of the number of tennis balls ordered. Select from the list below.

    +   −   ÷   ×

    10   20   30   40   200   300   400   600   800   1,200

    ☐ ☐ ☐ = ☐

    Part B: Is the estimated number of tennis balls ordered greater than or less than the actual number? Explain how you know.

13. Mr. Holt has 26 students in his class. He passed a bin of markers around the room 11 times. The first 10 times that the bin came around, every student selected a marker. The last time, only some of the students selected a marker. Others did not get to select a marker because the bin was empty before it got to them. Select **all** choices that could be the original number of markers in the bin.

    ☐ 26   ☐ 47   ☐ 261   ☐ 280

    ☐ 37   ☐ 260   ☐ 272   ☐ 286

14. In a cycling relay race, each team member rides for 16 kilometers. Team Rowdy Riders has 11 team members. On race day, all of the Rowdy Riders carpooled to the race. They rode in the race, and then rode their bikes to a lake to cool off. Finally, they rode back to the race site. The lake is 3 kilometers from the race site. For what total distance did the Rowdy Riders ride their bikes that day?

    A. 143 kilometers   C. 209 kilometers

    B. 176 kilometers   D. 242 kilometers

15. Corny's Roasted Corn stand bought 35 dozen ears of corn for the fair. Corny's sold 130 ears on the first day of the fair, and 162 on the second day. Corny's sold the rest of the ears on the third day. Indicate whether each statement is true or false.

True False

☐ ☐ More ears of corn were sold on the third day of the fair than on the first or second day.

☐ ☐ Corny's started the fair with more than 400 ears of corn.

☐ ☐ The equation $35 \times 12 - 130 + 162 = c$ can be used to figure out the number of ears of corn sold on the third day.

☐ ☐ The expression 35×12 can be used to figure out the total number of ears of corn sold at the fair.

16. A group of 74 students and 8 teachers order tickets for a theater festival. Student tickets cost $12 each, and teacher tickets cost $15 each. There is a fee of $5 for each group order.

    Part A: Complete the equation so that it shows the total amount, t, that the group paid for the order.

    74 ☐ ☐ + ☐ × ☐ + ☐ = t

    Part B: Solve the equation. Show your work and state the answer.

NAME _____   DATE _____

Chapter 6 Test

SCORE _____

1. A charity sends out 9 volunteers to conduct a survey of 270 households. Each volunteer will survey the same number of households. Explain how to use a basic fact and place value to find the number of households each volunteer will survey.

   [   ]

2. Gianna is going on a cross-country trip with her family. They plan to drive 2,046 miles over 7 days. Gianna estimates that her family will drive about 300 miles per day. Select **each** true statement about Gianna's calculation.

   > **ONLINE TESTING**
   > On the actual test, you might click to select options. In this book, you will use a pencil to shade the boxes.

   ☐ Gianna rounded to the nearest hundred before dividing.

   ☐ Gianna used compatible numbers to estimate.

   ☐ Gianna's estimation can be checked with the product 300 × 7.

   ☐ Gianna's estimation can be checked by adding 46 to 2,000.

   ☐ Gianna's estimate is greater than the actual distance the family will drive if they travel the same number of miles each day.

3. The cafeteria manager must divide 63 bagels equally among 3 different fourth grade classes. The model below shows how he did this.

   Part A: Complete the division sentence shown by the model.

   Part B: What does the answer to the division sentence mean in terms of the problem situation?

   [   ]

Grade 4 • Chapter 6 Divide by a One-Digit Number   83

4. Ms. Simmons spent a total of $520 at the animal clinic for her dog and three cats. She spent the same amount on each pet.

   Part A: Model the total amount Ms. Simmons spent using the least number of base-ten blocks possible.

   Part B: Show how to use the base-ten blocks to solve the problem. You may trade for different base-ten blocks, if necessary.

5. Mark "Y" for **each** division statement that has a remainder and "N" for **each** that does not.

Y	N	
☐	☐	132 ÷ 10
☐	☐	65 ÷ 5
☐	☐	231 ÷ 8
☐	☐	56 ÷ 7
☐	☐	136 ÷ 4

6. Shawn finds 5)1,545 and gets 39. He checks it and gets 39 × 5 = 195. State which calculation was wrong: the division or the multiplication. Then use place value to explain the mistake that Shawn made.

7. The fourth graders at Mountainview Elementary are sorting canned food donations into boxes for a community kitchen. The table shows the number of each type of donation.

Type of Food	Number of Cans
Peaches	32
Pears	45
Peanut butter	37
Black or pinto beans	66

Part A: How many boxes can be packed if each box must contain 8 cans that contain a protein food (beans or peanut butter)? Show your work, and tell how you know whether there will be any protein food cans left over.

Part B: The cans of fruit will be divided among the number of boxes determined in **Part A**. Explain how to find the number of cans of fruit that will go into each box.

Grade 4 • Chapter 6 Divide by a One-Digit Number 85

8. Write the missing digits that complete the division problem.

    ```
       □□4R3
    6)134□
    ```

    ONLINE TESTING
    On the actual test, you might use a keyboard to type in answers. In this book, you will write the answers.

9. A company has 593 large and 767 small packages. The company has 20 delivery trucks. Each truck will deliver the same number of packages to a shipping port. Use some of the numbers and symbols shown to write an equation that can be used to find p, the number of packages that each truck will carry.

    ( )   +   −   ÷   ×   =   593   767   20 p

10. Jennae plans to buy one of the computers whose price is shown. She wants to divide the cost of the computer into 4 payments. Write the prices in the correct bin according to whether each computer can be paid for with 4 payments of the same exact whole dollar amount.

4 Equal Whole Dollar Payments	4 Unequal Whole Dollar Payments

11. Jon has 533 pennies. He wants to put the same number of coins in each of 6 piggy banks. He uses the division statement 540 ÷ 6 to estimate that he will put about 90 pennies in each bank. Ellie states that Jon estimated wrong because 533 does not round to 540. Is Ellie correct? Explain.

86 Grade 4 · Chapter 6 Divide by a One-Digit Number

12. Four third grade classes, 3 fourth grade classes, and 3 fifth grade classes are selling wrapping paper to raise money. Each class will sell the same number of rolls of each type of wrapping paper. The numbers of each type are shown in the table.

Wrapping Paper Fundraiser			
Type of Paper	Number of Rolls	Rolls per Class	Rolls Left
Valentine's Day	37		
Birthday	223		
Baby Shower	50		
Wedding	185		
Winter Holiday	360		

Part A: How can you tell, without dividing, whether the number of rolls of each type can be divided evenly among the classes?

Part B: Find how many of each type of paper each class will sell. State how many rolls of each type, if any, will be left over. Write the values in the table.

13. Mr. Jeong wants to figure out how much money he can spend on his garden center expenses for each of the first 7 weeks after it opens. He uses the problem $7\overline{)4,340}$. Determine whether each statement is true or false.

True False

☐ ☐ His problem has a remainder.

☐ ☐ He can spend $620 each week.

☐ ☐ He can spend $627 each week.

☐ ☐ He starts with a total of $4,340 for expenses.

☐ ☐ He starts with a total of $30,380 for expenses.

14. Sani is in a group of 4 art students. He wants to figure out how many sheets of construction paper each student in his group should get. He has 234 sheets. Sani will check his work. Write the numbers 1–3 next to the correct three steps below to show the order in which Sani should perform them.

    _____ Add 2 to 232.

    _____ Divide 234 by 4.

    _____ Divide 58 by 4.

    _____ Multiply 58 by 4.

    _____ Subtract 2 from 234.

15. A company orders 337 desks. Four of them are used for the reception area. The rest of the desks will go on the company's 3 floors of offices. Each floor will get the same number of desks.

    Part A: Write an equation that can be used to find *d*, the number of desks that will go on each floor.

    Part B: Think about the numbers in the division problem. Explain how you can solve the problem using place value and mental math. Then state the answer.

16. Eduardo has a school supply allowance of $50. He bought 2 packs of pencils for $2 each, a pencil sharpener for $1, a pack of markers for $9, and a backpack for $25.

    Part A: How many $3.00 notebooks can Eduardo buy with the remaining money? Show your work.

    Part B: Will Eduardo have any money left over? If so, how much? Explain how you know.

Chapter 7 Test

1. Look at the number list.

   Part A: Extend the pattern.

   48, 46, 51, 49, 54, 52, _____, _____

   Part B: Complete the statement.

   The pattern is _____.

2. Aaron is making a pattern with toothpicks. Each day, he adds to the pattern, as shown in the picture. Select **all** true statements below about the pattern.

   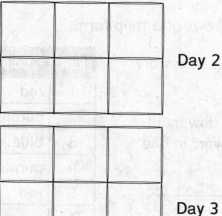

   Day 1

   Day 2

   Day 3

   ONLINE TESTING
   On the actual test, you might click to select options. In this book, you will use a pencil to shade the correct boxes.

   ☐ It is a growing pattern.

   ☐ A rule for the pattern is "Add one rectangle."

   ☐ It will have exactly 24 squares on the 8th day.

   ☐ It will have exactly 70 toothpicks on the 10th day.

   ☐ Aaron can figure out the number of squares it will have on the 12th day without actually continuing the pattern.

Grade 4 · Chapter 7 Patterns and Sequences

3. Carrie is decorating the front of her notebook with stickers in the pattern shown.

Part A: Draw the next three stickers that Carrie will use.

Part B: Is this a growing pattern or a repeating pattern? Explain how you know.

4. The table shows the colors of flowers in several rows of a tulip farm.

Part A: Complete the table. Then circle the rows necessary to show the pattern unit.

Part B: Compare the number of rows of purple flowers to the number of rows of red and rows of blue flowers in one pattern unit.

Row	Color
1	red
2	purple
3	blue
4	purple
5	red
6	
7	
8	

5. Ms. Chopra listed the first 10 multiples of 8 on the board. Select **all** correct statements.

☐ All of the numbers on the list are even.

☐ The ones digits in the list repeated the following pattern: 8, 6, 4, 2, 0

☐ The sum of the digits in each number in the list is divisible by 4.

☐ Each number in the list is divisible by 4.

☐ Each number in the list is divisible by 16.

6. Lena picks 3 tomatoes from her garden on Monday, 3 on Tuesday, 6 on Wednesday, 12 on Thursday, and 24 on Friday. She notices that the number she picks each day after Monday is the sum of the numbers she picked on all previous days of the week. What is another pattern that describes the number picked on any day after Monday?

7. Each day for 5 days, Tolu uses 9 sheets of notebook paper. Which pattern could be a list of the number of sheets he has **left** at the end of each day?

   A. 9, 18, 27, 36, 45

   B. 80, 71, 63, 54, 45

   C. 10, 19, 28, 37, 46

   D. 65, 56, 47, 38, 29

8. Kimiko is training for a long distance race. She sets a goal to run 5 kilometers the first week. During the first week, she runs 6 kilometers farther than her goal. Each week, she runs 6 kilometers farther than she ran the previous week.

   Part A: How far does she run during the 8th week?

   Part B: Complete the equation with the values shown so that it can be used to find the distance, k, that Kimiko runs during week w.

   $$5 \quad 6 \quad K \quad w$$

   $$\square = \square \times \square + \square$$

9. The manager of a dinner theater uses the equation shown to figure out the amount of money, m, the theater earns when p people order a dinner and a performance. What is the value of m when $p = 18$? Show your work.

   $$(8 + 12) \times p = m$$

10. Look at the table.

Input (r)	4	8	12	16
Output (v)	12	24	36	48

Which equation describes the pattern?

A. $r \times 3 = v$ C. $r + 8 = v$

B. $v \times 3 = r$ D. $v + 8 = r$

11. Use order of operations to determine the unknown that completes the equation.

$$(7 + 4) \times (6 - \boxed{}) = 44$$

12. Tim is finding output values for the equation $100 \div (a + 1) = t$. He states that the output, t, is 26 when the input, a, is 4.

    Part A: What is Tim's mistake?

    Part B: Rewrite the equation so that Tim's value is correct.

13. A pattern is generated using this rule: start with the number 81 as the first term and divide by 3. Enter numbers into the boxes to complete the table.

Term	Number
First	81
Second	
Third	
Fourth	
Fifth	

14. A new café sells 22 cups of tea on the first day of the week. It sells 30 on the second day, and 38 on the third day. This pattern continues for the rest of the week. Indicate whether **each** statement is true or false.

    True   False

    ☐   ☐   The number of cups of tea sold by the café each day is always a multiple of 8.

    ☐   ☐   The rule for the number of cups of tea sold the next day is "add 8".

    ☐   ☐   The total number of cups of tea sold by the café during the week is even.

    ☐   ☐   The increase in the number of cups of tea sold each day forms a pattern.

15. A store clerk arranges groups of oranges in the pattern shown. Draw the next group in the pattern.

16. Extend the pattern.

17. In the table shown, each rule's input is the previous rule's output.

   Part A: Complete the table by writing in the missing values.

Input	Add 4	Subtract 2	Multiply by 3
5	9	7	
6			
	14	12	
		5	15

   Part B: If n is the input, write an equation that can be used to find q, the output.

18. The table shows the number of gallons of gas used by a truck driver on different trips and the amount paid for gas.

Input (gallons of gas)	24	203	89	327
Output (amount paid for gas)	$96	$812	$356	$1,308

   Based on the table, how is the output obtained from the input?

   A. Add 4.

   B. Multiply by 4.

   C. Add 72.

   D. Multiply by 48.

19. How many rings will be in the 12th figure in the pattern shown?

Chapter 8 Test

1. Roscoe arranged 80 square tiles in several rectangular patterns. Each pattern had more than 2 rows and had no tiles left over.

   Part A: Write the dimensions of the rectangular patterns Roscoe could use.

   Dimensions: ☐

   Part B: Roscoe wants to make a square pattern out of his tiles. If he makes the largest square possible, what will the dimensions of the square be? How many tiles will be left over? Draw and label Roscoe's square.

   Dimensions of square: ☐

2. A tech website ranked four printers with respect to printing speed. The table shows the amount of time each printer took to print one page.

   Part A: Rank the printers from fastest to slowest on the table.

Printer	A	B	C	D	E
Time to print 1 page (minutes)	$\frac{1}{6}$	$\frac{3}{8}$	$\frac{5}{12}$	$\frac{1}{4}$	$\frac{5}{24}$
Rank					

   Part B: Where would a printer that took $\frac{2}{5}$ of a minute to print a page rank on the table?

3. The City Council wants to pass a law that bans the use of plastic shopping bags. The law states that at least $\frac{5}{8}$ of the council must vote yes for a law to pass. The Council has 40 members. Select true or false for each statement.

True   False

☐   ☐   The measure will pass with 24 yes votes.

☐   ☐   The measure will pass if $\frac{3}{5}$ of the members vote yes.

☐   ☐   The measure will pass with 25 yes votes.

☐   ☐   The measure will pass if $\frac{15}{20}$ of the members vote yes.

4. Monica is cutting slats to make a butcher block. She cut 11 slats. The end of each slat is a square that is $\frac{1}{4}$ foot on each side.

Part A: If Monica stacks the slats on their long sides, how tall will the butcher block be?

Part B: Draw and label Monica's butcher block.

5. A 1-mile horse race is measured in furlongs. There are exactly 8 furlongs in 1 mile, so each furlong is $\frac{1}{8}$ of a mile. If a horse has completed 6 furlongs in a 1-mile race, how far has it run? Select the correct distances.

☐ $\frac{3}{4}$ mile   ☐ 6 miles

☐ $\frac{3}{4}$ furlong   ☐ $\frac{6}{8}$ mile

96 Grade 4 • Chapter 8 Fractions

6. Jadamian claims that half of the odd numbers that are less than 100 are prime numbers. Is Jadamian correct? Justify your answer.

7. At Bear Lake, Ranger Dina takes care of reserving camp sites. So far $\frac{1}{2}$ of the sites have been reserved for next week.

ONLINE TESTING
On the actual test, you might be asked to drag objects to their correct location. In this book, you will be asked to write or draw the object.

Part A: Dina has told all of the campers that their site does not share a side with another reserved site. Is Dina correct? Color in sites to justify your answer.

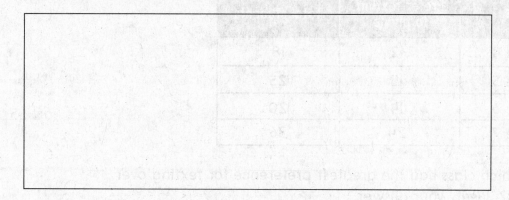
Bear Lake Campsites

Part B: Three more campers have now reserved sites. How will that change Dina's statement that no one will have a site that shares a side with another reserved site? Color in sites to justify your answer.

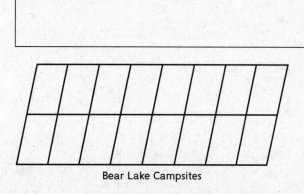
Bear Lake Campsites

Grade 4 • Chapter 8 Fractions 97

8. Rajiv conducted a survey of students' attitudes toward texting and phone calls. The results are shown in the table.

Class	Prefer texting to phone	Total students
Grade 4	12	18
Grade 5	15	25
Grade 6	14	20
Grade 7	24	36

Part A: Which class had the greatest preference for texting over phone calls? Justify your answer.

Part B: The eighth grade data for 30 students showed that eighth graders had the same preference for texting as sixth graders. How many eighth graders preferred texting over phone calls?

9. Batting records of baseball players were compared.

   - Diaz was a better hitter than Williams.
   - Diaz was not as good of a hitter as Cabrera.

   Each number shows the fraction of the time that each player got a hit. Which of the following records is possible for the three hitters?

   ☐ Diaz: $\frac{1}{3}$, Williams: $\frac{3}{10}$, Cabrera: $\frac{2}{5}$

   ☐ Diaz: $\frac{4}{10}$, Williams: $\frac{3}{10}$, Cabrera: $\frac{2}{5}$

   ☐ Diaz: $\frac{3}{8}$, Williams: $\frac{1}{3}$, Cabrera: $\frac{4}{9}$

   ☐ Diaz: $\frac{1}{4}$, Williams: $\frac{1}{3}$, Cabrera: $\frac{3}{8}$

10. Jana is building a wooden table and needs to measure a peg to fit into a hole. Jana knows that the hole is between $3\frac{3}{4}$ and $3\frac{9}{10}$ inches in width, but does not have a tape measure to find the exact width. Which of the following could be the actual width of the hole? Select all that apply.

☐ $3\frac{7}{10}$ inches ☐ $3\frac{5}{6}$ inches

☐ $3\frac{4}{5}$ inches ☐ $3\frac{11}{12}$ inches

11. Moncef, the owner of a software start-up, wants his 48 workers to split up into creative teams.

    • Each team must have more than 2 members.
    • Each team must have the same number of members.

    How many different arrangements of teams can the workers form? List them all in the box below.

12. Solve this riddle.

    Part A: I am a composite number between 50 and 60. Both of the factors that form me are prime numbers themselves. Which numbers might I be? Circle the possible numbers.

    50, 51, 52, 53, 54, 55, 56, 57, 58, 59, 60

    Part B: I am one of the numbers from Part A above. The sum of my factors equals a composite number that is a square of another number. Which number am I? Explain.

13. The box office at the Crystal Theater has sold three-fourths of its seats for Saturday's show. The seat map shown has not been updated to show that three-fourths of the available seats have been sold. Draw X's in seats to update the seat map and show exactly three-fourths of the seats sold.

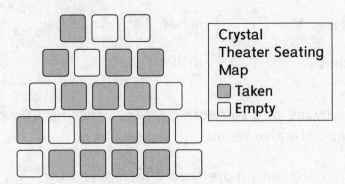

14. Determine whether each statement is true or false.

   True False
   ☐ ☐ $\frac{9}{12}$ is equivalent to $\frac{15}{20}$.
   ☐ ☐ $\frac{9}{12}$ is the simplest form of $\frac{15}{20}$.
   ☐ ☐ $\frac{9}{12}$ is equivalent to $\frac{18}{24}$.
   ☐ ☐ $\frac{3}{4}$ is the simplest form of $\frac{15}{20}$.

15. Roy planted his garden as shown below. Rank Roy's garden from greatest to least with respect to the amount of land each item had.

   - $\frac{2}{5}$ of the garden had tomatoes
   - $\frac{1}{12}$ of the garden had peppers
   - $\frac{1}{6}$ of the garden had lettuce
   - $\frac{2}{10}$ of the garden had herbs
   - $\frac{3}{20}$ of the garden had flowers

100 Grade 4 • Chapter 8 Fractions

Chapter 9 Test

1. Write an addition sentence for each model. Then find the sum.

   Part A:

   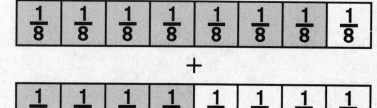

   Part B:

   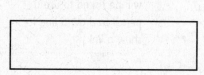

   Part C:

   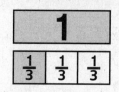 +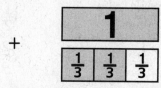

2. Which of the following is equivalent to $\frac{1}{2}$? Select all that apply.

   ☐ $\frac{5}{8} - \frac{1}{8}$

   ☐ $\frac{5}{16} + \frac{3}{16}$

   ☐ $\frac{11}{20} + \frac{3}{20}$

   ☐ $\frac{17}{20} - \frac{9}{20}$

Grade 4 · Chapter 9 Operations with Fractions

3. Sasha and Ali caught fish and weighed them before throwing them back into the water. Sasha's fish weighed $2\frac{2}{5}$ pounds. Ali's fish weighed $1\frac{4}{5}$ pounds. Draw a model that shows the total weight of both fish.

4. Draw a line between each pair of equivalent expressions.

$8 \times \frac{1}{3}$　　　　$5 \times \frac{2}{7}$

$\frac{10}{7}$　　　　　$4 \times \frac{2}{5}$

$8 \times \frac{1}{5}$　　　　$1\frac{1}{2}$

$9 \times \frac{1}{6}$　　　　$2\frac{2}{3}$

ONLINE TESTING
On the actual test, you might be asked to click on objects to link them together. In this book, you will be asked to use a pencil to draw a line to show a link.

5. Rosa wrote these addition and subtraction sentences. Describe each mistake that Rosa committed. Correct the mistake. If the sentence is correct, write "Correct."

$\frac{3}{5} + \frac{2}{5} = \frac{5}{10}$

$\frac{5}{12} - \frac{1}{12} = \frac{1}{3}$

$\frac{3}{10} + \frac{3}{10} = \frac{2}{5}$

$\frac{11}{20} - \frac{3}{20} = \frac{2}{5}$

6. The water tank for Camp Utmost is almost full. The head camp counselor, needs to file a report with the water department if the tank gets down to half full.

Part A: What fraction of the tank is now filled? Explain how you found your answer.

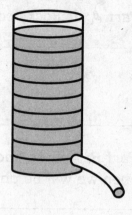

Part B: How much more water needs to be drained out of the tank to cause the counselor to file a report with the water department?

Part C: The counselor needs to put the camp on emergency alert if the level gets down to $\frac{1}{5}$ full. From its starting level, how much more water would need to be drained out of the tank to cause an emergency alert to be issued?

7. Fill in the missing items to subtract $2\frac{3}{4}$ from $3\frac{1}{4}$. Simplify your answer.

$$3\frac{1}{4} = \frac{4}{4} + \frac{4}{4} + \boxed{} + \boxed{} = \frac{\boxed{}}{4} = \frac{\boxed{}}{4}$$

$$-2\frac{3}{4} = \boxed{} \qquad = \frac{\boxed{}}{4} = \frac{\boxed{}}{4}$$

$$= \boxed{}$$

8. Andrea and Rocky mixed $\frac{4}{6}$ cup of whole wheat flour with $\frac{8}{12}$ cup of rye flour. To find out how much flour they have in all, Rocky said, "We can just add the numerators of the fractions."

   Part A: Is Rocky correct? Explain your answer.

   Part B: Andrea had an idea. "Why don't we simplify both fractions. Then we will be able to add them." Is Andrea correct?

   Part C: Simplify the two fractions. Are you now able to add the two fractions? If so, what sum do you get? If not, explain why the two fractions could not be added.

9. Draw a model to show the multiplication of $3 \times \frac{5}{8}$.

10. Select true or false for each statement.

　　True　False
　　☐　☐　$2\frac{4}{9} + 4\frac{8}{9} = 6\frac{1}{3}$
　　☐　☐　$4\frac{4}{9} - 2\frac{8}{9} = 1\frac{5}{9}$
　　☐　☐　$6\frac{7}{10} + 6\frac{7}{10} = 13\frac{2}{5}$
　　☐　☐　$7\frac{7}{10} - 6\frac{9}{10} = 1\frac{8}{10}$

11. Draw a line between each pair of equivalent expressions.

$3\frac{2}{5} + 2\frac{2}{5}$ 　　　　　　　　　$5\frac{5}{8} + 3\frac{7}{8}$

4 　　　　　　　　　　　　　　$\frac{29}{5}$

$3\frac{5}{6} + 3\frac{5}{6}$ 　　　　　　　　　$2\frac{5}{12} + 1\frac{7}{12}$

$\frac{45}{8} + \frac{31}{8}$ 　　　　　　　　　$7\frac{2}{3}$

12. Joachim started the day with $3\frac{2}{5}$ gallons of paint. He used $1\frac{4}{5}$ gallons to paint the garage door. Then Joachim painted the shed. When he was done painting, Joachim had $\frac{2}{5}$ gallon of paint left. How much paint did Joachim use for the shed? Show how you got your answer.

13. Use the models to find the sum of the two quantities. Fill in the boxes. Shade the circles and draw new circles if necessary.

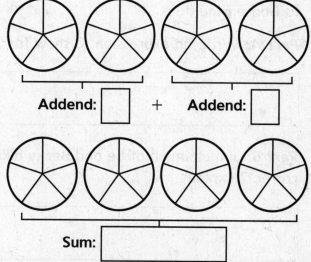

Addend: ☐ + Addend: ☐

Sum: ☐

14. Mona swam back and forth across the Narrow River 8 times. The Narrow River measures $\frac{1}{5}$ of a mile across. Elise swam $\frac{2}{3}$ of the way across the Wide River that measures 5 miles across.

   Part A: Which expressions show how far Mona swam? Select all that apply.

   ☐ $8 \times \frac{1}{5}$
   ☐ $16 \times \frac{1}{5}$
   ☐ $\frac{2}{5} \times 8$
   ☐ $\frac{2}{5} \times 16$

   Part B: Which expressions show how far Elise swam? Select all that apply.

   ☐ $\frac{2}{3} \times 5$
   ☐ $10 \times \frac{1}{3}$
   ☐ $20 \times \frac{1}{3}$
   ☐ $10 \times \frac{2}{3}$

   Part C: Who swam farther, Mona or Elise? Explain your answer.

15. The hike from Ducktown to Goose Mountain is $7\frac{4}{5}$ miles. Jim and Becky hiked $1\frac{2}{5}$ miles toward Goose Mountain. Then Jim forgot his water so they hiked back to Ducktown. Then Jim and Becky hiked all of the way to Goose Mountain.

   Part A: Write an expression to show far Jim and Becky hiked in all.

   Part B: Jim claims that he and Becky hiked more than 10 miles. Is Jim correct? Explain.

Chapter 10 Test

1. Mindy bought a blue marker for 0.93 dollars. Which coins should she take out to pay for the marker? Draw quarters, dimes, nickels, and pennies in the box. Label each coin.

> **ONLINE TESTING**
> On the actual test, you may drag coins into the box to show your total. In this book, you will use a pencil to draw in and label the coins that you use.

2. Decimal models and a number line are shown.

   Part A: Draw a line from each model to its location on the number line.

   Part B: Write the numbers in **Part A** from least to greatest.

Grade 4 · Chapter 10 Fractions and Decimals **107**

3. Marques claimed he could write a number or model that is equivalent to 0.70 in six different ways. Is Marques correct? Explain your answer.

4. Shondra made this model to show the sum of $\frac{6}{10}$ and $\frac{3}{100}$.

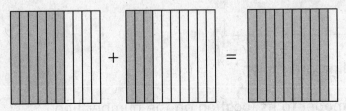

Part A: Is Shondra's model correct? Explain.

Part B: Draw a correct model for sum of the two fractions.

Part C: What is the sum $\frac{6}{10} + \frac{3}{100}$ in decimal form?

5. A jeweler melted down a silver nugget and an old necklace to obtain two batches of pure silver. The silver from the nugget weighed 0.7 ounces. The silver from the necklace weighed $\frac{70}{100}$ of an ounce. Which piece had greater value? Explain.

6. Jay's recipe for making tortillas calls for 0.6 pounds of corn meal and 0.19 pounds of vegetable oil.

   Part A: Which ingredient is greater in weight? Explain.

   Part B: Jay adds both ingredients to a bowl. How much do the ingredients in the bowl weigh? Color in the model to show your answer. Write the total in decimal and fraction form.

   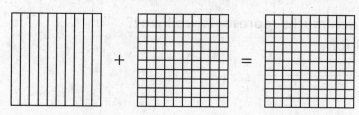

7. Yasmina is sewing a jacket. For the sleeves Yasmina cuts a piece of cloth that is $\frac{7}{10}$ of a meter in length and 0.2 meter in width. For the front panel Jasmine cuts a piece of cloth that is $\frac{65}{100}$ meter in length and 0.2 meter in width.

   Part A: Yasmina compares the lengths of the two pieces. To find out which piece is longer, what information does Yasmina need?

Part B: What do you know about the widths of the pieces?

Part C: Which piece of cloth is longer? Explain.

8. Belinda made this model of her garden. Each shaded square shows where flowers are planted. Non-shaded squares show the walkways in Belinda's garden.

   Part A: What part of the garden has flowers? Write your answer in decimal form.

   Part B: What part of the garden marks represents walkway? Write your answer in decimal form.

9. Use the model to show a different arrangement of flowers for Belinda's garden in problem 8. The arrangement should have the same amount of space for flowers as Belinda's garden. Explain why your model is equivalent to Belinda's model.

10. At the tech store, Ky saw a case for her phone that cost $4.27. Fill in the table to show the money Ky needs for the phone case.

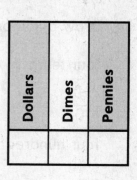

11. Quinn collected $\frac{26}{100}$ of a kilogram of raspberries. She worked from 9:30 A.M to 1:00 P.M. Mitch collected 0.4 kilograms of raspberries. With the information given, which of the following problems can you solve? Select all that apply.

    ☐ When did Mitch stop collecting raspberries?
    ☐ How many kilograms did Mitch and Quinn pick together?
    ☐ How many kilograms did Quinn pick by 12 noon?
    ☐ Who picked more raspberries, Quinn or Mitch?

12. Arrange these numbers in the box from least to greatest.

    0.22 $\frac{2}{10}$ 0.02 $\frac{23}{100}$ 0.21

13. Noah ran $\frac{3}{10}$ of a mile to Unger Park. In the park, Noah walked for 0.9 mile with his friend Oswaldo. Finally, Noah and Oswaldo ran 0.48 mile to Oswaldo's house.

    Part A: How far did Noah run in all?

    Part B: Did Noah run farther than he walked? Explain.

14. Draw a line between each pair of equivalent expressions.

    four tenths   0.04

    $\frac{77}{100}$ 0.70

    four hundredths   0.40

    0.7   0.77

15. Select true or false for each statement.

    True False
    ☐   ☐ $0.1 + \frac{1}{10} = 1.1$
    ☐   ☐ $0.2 + \frac{2}{10} = 0.22$
    ☐   ☐ $0.03 + \frac{3}{10} = 0.33$
    ☐   ☐ $0.04 + \frac{4}{100} = 0.08$

16. Danny says he can show this amount of money using just a single type of coin. Which coin does he need to use and how many of these coins will he need? What is the total number of cents?

112 Grade 4 · Chapter 10 Fractions and Decimals

NAME

DATE

SCORE

Chapter 11 Test

1. An experimental car that runs on solar power weighs $2\frac{1}{4}$ tons. How many pounds does this car weigh?

2. Compare the horizontal line segments for the two figures.

   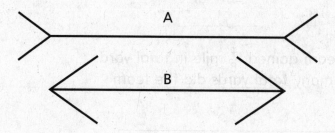

   Part A: Which horizontal line is greater in length? Make an estimate.

   Part B: Measure each horizontal line segment to the nearest $\frac{1}{4}$ inch. Which figure is greater in length?

   Part C: How accurate was your estimate? Explain.

3. Elle poured 1 quart, 1 pint, and 1 cup full of water into an empty bucket. How many additional cups of water will it take to fill the 1-gallon bucket?

Grade 4 • Chapter 11 Customary Measurement 113

4. Raymond's recipe for tortilla soup calls for filling this cook pot with broth, black beans, and tomatoes. Raymond does not know how much the pot holds. What would be a good estimate for the capacity of the pot? Select all that apply.

☐ 9 gallons
☐ $2\frac{1}{2}$ quarts
☐ 5 pints
☐ 80 fluid ounces

5. Coach Ruiz tells says his football team gained $1\frac{1}{8}$ mile in total yards from scrimmage this season. How many total yards did the team gain?

6. Draw a line between each pair of equivalent quantities.

8 cups	5 cups
$2\frac{1}{2}$ gallons	1 pint + 1 cup
40 fluid ounces	10 quarts
24 fluid ounces	4 pints

7. Ms. Stephanos bought a house with a 30-year mortgage. She will be making the same payment each month for the next 30 years.

Part A: How many monthly payments will Ms. Stephanos make? Explain.

Part B: How many weeks will be covered over the 30 year period? Explain.

8. Wilson measured the speed of his tennis serve and rounded each speed to the nearest 5 miles per hour.

45 mph	50 mph	55 mph	60 mph	65 mph	70 mph
1	3	5	4	0	1

Part A: To make a line plot using Wilson's data, you first need to make a scale. Which high and low values and intervals should you choose for the scale? Draw in the values and intervals on the scale.

Part B: Use the data to draw in the values for the line graph.

Part C: What was the speed of a typical serve for Wilson? Explain.

9. Boris's skateboard helmet weighs $1\frac{3}{8}$ pounds. Marisa's helmet weighs $20\frac{1}{4}$ ounces.

Part A: Whose helmet weighs more? Explain how you know.

Part B: How much less does the lighter helmet weigh?

Grade 4 • Chapter 11 Customary Measurement

10. Sandy ran 3 laps around a $\frac{1}{4}$-mile track. Which of the following are equal to how far Sandy ran? Select all that apply.

☐ 1,320 yards
☐ 4,000 feet
☐ 3,960 feet
☐ 3,960 inches

11. Ephraim estimated the distance that he hit a golf ball. Place a value from the list below inside each box.

ONLINE TESTING
On the actual test, you may drag items into boxes. In this book, you will write the values in the boxes using a pencil.

| 176 | 6,336 | $\frac{1}{10}$ | 528 |

feet	yards	inches	miles

12. A blue whale's tongue weighs about $3\frac{1}{4}$ tons. A full-grown hippopotamus weighs 6,150 pounds.

Part A: Which weighs more, the blue whale's tongue or the hippopotamus? Explain how you know.

Part B: How much more does the heavier item in **Part A** weigh?

116 Grade 4 · Chapter 11 Customary Measurement

13. Write the measurements from the list in order from least to greatest.

| $\frac{1}{4}$ quart | $\frac{1}{8}$ gallon | 3 fluid ounces | $1\frac{1}{2}$ pint | $\frac{1}{4}$ cup |

14. Davin ran the marathon in $4\frac{3}{5}$ hours. Melba ran the marathon in 266 minutes. After she crossed the finish line, Melba waited for Davin to arrive. How many seconds did she wait? Explain.

15. LeVar has 9 coins in his pocket that are worth a total of 86 cents. LeVar has at least one of each of the following kinds of coin: penny, nickel, dime, and quarter. LeVar has more nickels than dimes. What coins does LeVar have? Explain.

Grade 4 · Chapter 11 Customary Measurement 117

16. Select true or false for each statement.

    True False
    ☐   ☐   7 cups > 2 quarts
    ☐   ☐   2 quarts > 1 pint + 42 fluid ounces
    ☐   ☐ $2\frac{1}{8}$ gallon = 17 pints
    ☐   ☐ $\frac{1}{8}$ quart < $\frac{3}{4}$ cup

17. Nilsa and her classmates collected donations to get new computers for her school. Donations were made in increments of $10. Nilsa made this line graph of the data, showing donations in dollars.

    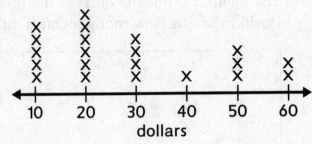

    Part A: What was the most common amount that people donated?

    Part B: Which donation amount produced the greatest donation total? What was the total donated at that amount?

18. Franklin wants to use a bowling ball that weighs no more than 224 ounces. The label on the box for a new bowling ball shows that the ball weighs 12 pounds 14 ounces. Is this a ball that Franklin could use? Explain.

NAME .... DATE ....

Chapter 12 Test

SCORE ....

1. Select all of the items that would have a volume of about 0.2 L.

   ☐ cup of hot tea

   ☐ jug of ice tea

   ☐ bottle of lotion

   ☐ sip of soup

   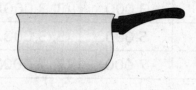

2. Jerrilyn's wrist watch weighs 0.15 kilograms. Josh's watch weighs 145 grams.

   Part A: Whose watch is heavier? Explain.

   Part B: In grams, how much greater in weight is the heavier watch?

3. A plan for a fountain is shown. The diameter of the outer circle is 8 meters. The distance between the outer circle and the shaded inner circle measures 2.5 meters. What is the diameter of the shaded inner circle in centimeters? Show your work.

   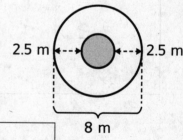

4. In Science lab, Regan needs to fill a tank with 2.25 liters of distilled water. How many times will Regan need to refill a 250 mL beaker with water in order to get the 2.25 liters that she needs?

Grade 4 • Chapter 12 Metric Measurement 119

5. Which units would this conversion table fit? Select all that apply.

?	?	?
3	3,000	(3, 3,000)
5	5,000	(5, 5,000)
7	7,000	(7, 7,000)

☐ meters and centimeters

☐ liters and milliliters

☐ kilograms and grams

☐ centimeters and kilometers

6. Tracee has 3 songs on her new playlist with titles "Always," "Before I Leave," and "Come Home." On her playlist, Tracee abbreviates the songs as A, B, and C. In how many different orders can Tracee play the 3 songs? Show your work.

7. Monty measured the distance across the Juniper Park swimming pool. To show this distance, place a value from the list below inside each box.

ONLINE TESTING
On the actual test, you may drag items into boxes. In this book, you will write the values in the boxes.

cm	km	mm	m

3,200   32   0.032   32,000

120 Grade 4 · Chapter 12 Metric Measurement

8. Fresh mozzarella cheese from Antonio's Market costs $4.00 per kilogram.

   Part A: Roxanne paid $8.00 for her mozzarella. Did she buy enough cheese to make one large pizza? Explain.

   Part B: Roxanne bought 2 large watermelons from Antonio's market. Are the watermelons likely to weigh more or less than the cheese she bought?

9. Draw a line between each pair of equivalent quantities.

250 mm	25,000 mm
250 cm	0.25 m
0.25 km	2.5 m
25 m	250 m

10. Select true or false for each statement.

    True False

    ☐ ☐   250 g > 2.5 kg

    ☐ ☐   3.1 L < 400 mL + 2 L

    ☐ ☐   150 mm > 5 cm

    ☐ ☐   0.4 km < 4,000 m

Grade 4 · Chapter 12 Metric Measurement

11. In darts, Alison scored 57 points on 8 throws. She hit each score on the board at least once.

    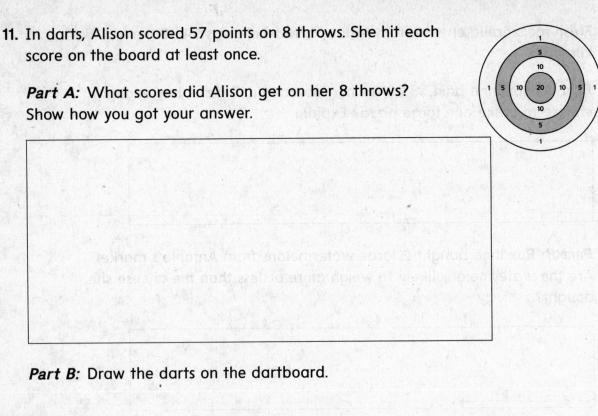

    Part A: What scores did Alison get on her 8 throws? Show how you got your answer.

    Part B: Draw the darts on the dartboard.

12. On a metric scale, Ira's fish weighed 1 kilogram. On a customary scale, the same fish weighed a little over 2 pounds.

    Part A: Ira caught a fish that weighed 4,000 grams. About how much would this fish weigh in pounds?

    Part B: Which would weigh more, a 10 kg fish or a fish that weighed 15 pounds? Explain.

13. At the county track meet, these were the top 5 finishers in the long jump event. Rank each finisher from first to fifth place.

Name	Distance	Place
Brian	4.08 m	
Shondra	403 cm	
Dave	399.9 cm	
Kayla	4.1 m	
Hernán	4,000 mm	

14. Rita is organizing the Hartley Park Treasure Hunt, in which students hunt for buried treasure in the park. Hartley Park is 0.1 kilometers wide and 100 m in length.

Part A: Rita's directions are shown. Circle the most likely choice of unit for each measurement.

1. Start at the main gate of the park. Walk 0.06 (millimeters, centimeters, meters, kilometers) north past the playground to the drinking fountain.

2. From the drinking fountain, walk 35 (millimeters, centimeters, meters, kilometers) through the playground west to the fence.

3. From the fence, walk 2,500 (millimeters, centimeters, meters, kilometers) south through the hopscotch court to the creek.

4. At the creek, turn and walk about 5 steps, or 3,000 (millimeters, centimeters, meters, kilometers) north. You will find the treasure there.

Part B: To find the treasure, what total distance do you need to walk? Show your work.

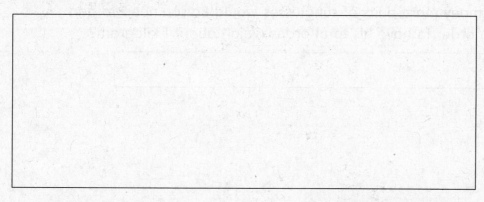

Grade 4 • Chapter 12 Metric Measurement

15. The figure is placed next to a centimeter ruler.

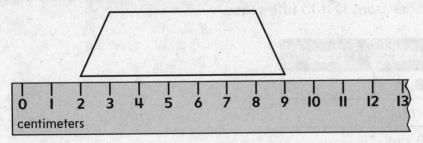

Part A: What is the approximate length of the bottom side of the figure? Select all that apply.

☐ 9 cm ☐ 70 mm

☐ 700 mm ☐ 0.07 m

Part B: What is the approximate length of the top side of the figure? Select all that apply.

☐ 50 mm ☐ 0.05 m

☐ 8 cm ☐ 0.5 cm

16. At the beach, Edison sells sunglasses and tubes of sunblock. He orders 15 pairs of sunglasses that weigh 26 grams each. He also orders 12 tubes of sunblock that weigh 43 grams each.

Part A: Is the total weight of Edison's order greater than or less than 1 kilogram? Explain.

Part B: How many more pairs of sunglasses would Edison need to add or subtract in order to have his total order weigh about 1 kilogram?

Chapter 13 Test

1. The rectangle has a length of 8 centimeters and an area of 24 square centimeters.

   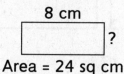
   Area = 24 sq cm

   Part A: What is the width of the rectangle?

   Part B: Draw in the square unit centimeters to show the area of the figure.

2. What is the next figure in this pattern? Draw the figure. What is the area of the figure?

3. The length of a rectangle is twice the width of the rectangle. The perimeter of the rectangle is 36 inches. Complete the table.

Length (in.)	Width (in.)	Perimeter (in.)	Area (sq. in.)

   ONLINE TESTING
   On the actual test, you may drag items into boxes. In this book, you will write the values in the boxes.

4. Draw a line between each pair of matching items.

   square feet   length × width

   area of rectangle   perimeter units

   inches   (length × 2) + (width × 2)

   perimeter of rectangle   area units

Grade 4 · Chapter 13 Perimeter and Area

5. Vanessa is an architect. She designed Goldfish Pond I for her client as a square.

   Part A: Find the perimeter of Goldfish Pond I.

   Part B: Vanessa changed Goldfish Pond I into Goldfish Pond 2. Pond 2 has a square cut-out that measures 4 feet on each side. What is the perimeter of Pond 2? Show your work.

   Part C: Compare the perimeters of Pond I and Pond 2. Why are they different? Explain.

6. Vanessa from Problem 5 made a sketch of Goldfish Pond H.

   Part A: Without knowing the size of the cut-outs, what can you say about the perimeter of Pond H compared to Pond I and Pond 2? Which pond will have the greatest perimeter? Which will have the least perimeter? Explain.

   Part B: Change Pond H so that its perimeter increases. Make a drawing.

126 Grade 4 · Chapter 13 Perimeter and Area

7. Carlos bought 28 meters of fencing to create a dog run for his dog. The dog run must be measured in whole meters.

   Part A: Fill in the table. Each dog run arrangement must use all of the fencing that Carlos bought.

Length (ft)	Width (ft)	Perimeter (ft)	Area (sq ft)
12		28	24
	3	28	
10			
			45
8			

   Part B: To give his dog the greatest area for play, which design should Carlos choose? Explain.

8. Carlos in Problem 7 created a dog run for Mr. Escobar using 40 feet of fencing. Mr. Escobar's dog run has 100 square feet of space. What are the dimensions of Mr. Escobar's dog run? Explain.

9. The figure is made of small squares that measure 3 cm on a side. What is the area of the shaded part of the figure?

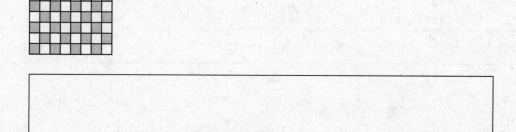

Grade 4 • Chapter 13 Perimeter and Area

10. Jazmine created a floor plan for her "dream" kitchen.

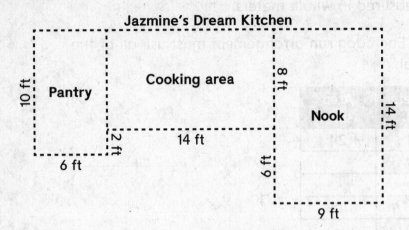

Part A: In order to find the area of her dream kitchen, Jazmine divided the kitchen into three areas: the Cooking Area, the Pantry, and the Nook. What is the total area of Jazmine's dream kitchen?

Part B: What is the perimeter of Jazmine's dream kitchen?

Part C: To check her perimeter answer, Jazmine found the sum of the perimeters of the three areas. Does the sum of the perimeters of the three areas agree with the perimeter you found in **Part B**? If so, explain why. If not, explain which value for the perimeter is correct.

11. Paul is putting in a 36 square foot model train course in his basement. The course must have a rectangular shape and be measured in whole meters. The course can have any length but it must be at least 2 meters in width.

    Part A: What are the possible dimensions for the course?

    [ ]

    Part B: Paul wants the trains to have the longest course possible and still take up only 36 square feet of space. Which dimensions for the course should he choose? Explain.

    [ ]

12. The swimming pool is surrounded by a 1-meter walkway. Compare the area of the walkway and the area of the pool. Which area is greater? By how much is it greater?

    [ ]

13. Select true or false for each statement about P, the perimeter of a figure.

    True   False

    ☐   ☐ $P = 4 \times$ side for all squares

    ☐   ☐ $P = 4 \times$ side for all rectangles

    ☐   ☐ $P = (2 \times \text{length}) + (2 \times \text{width})$ for all squares

    ☐   ☐ $P = (2 \times \text{length}) + (2 \times \text{width})$ for all rectangles

14. A rectangular indoor running track measures 60 yards by 52 yards. Giorgi ran 1,500 yards around the track. How many laps did he finish? Make a table to solve the problem.

Lap	Yards
1	
2	
3	

15. Rectangle A measures 6 cm by 4 cm. Rectangle B has the same area as Rectangle A and is 1 cm in width. Which of the following is true? Select all that apply.

☐ The perimeter of Rectangle A is greater.

☐ The perimeter of Rectangle B is greater.

☐ Rectangle B is greater in length than Rectangle A.

☐ Rectangle A is greater in length than Rectangle B.

16. Match the dimensions of each rectangle with the correct area and perimeter.

4 ft × 6 ft   P = 20 ft, A = 16 sq ft

5 ft × 5 ft   P = 20 ft, A = 24 sq ft

8 ft × 2 ft   P = 28 ft, A = 24 sq ft

12 ft × 2 ft   P = 20 ft, A = 25 sq ft

Chapter 14 Test

1. Consider the figure shown.

   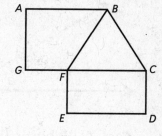

   Part A: Which line segments are parallel? Select all that apply.

   ☐ $\overline{AG}$ and $\overline{AB}$
   ☐ $\overline{FE}$ and $\overline{CD}$
   ☐ $\overline{FB}$ and $\overline{FC}$
   ☐ $\overline{AG}$ and $\overline{FE}$

   Part B: Which line segments are perpendicular? Select all that apply.

   ☐ $\overline{AG}$ and $\overline{AB}$
   ☐ $\overline{GC}$ and $\overline{CD}$
   ☐ $\overline{FB}$ and $\overline{GF}$
   ☐ $\overline{GF}$ and $\overline{FE}$

2. Draw a line between each pair of matching items.

obtuse angle	right angle
acute angle	$\frac{1}{6}$ turn
$\frac{1}{4}$ turn	180°
straight angle	greater than 90°

3. Draw an obtuse triangle with two equal acute angles.

Grade 4 • Chapter 14 Geometry

4. A billiard ball travels in the direction shown, bounces off a wall at angle *a* and rebounds at the exact same angle. The angle between the incoming path and outgoing path of the ball is 120°.

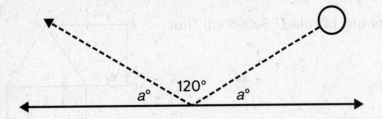

Part A: What is the sum of the measures of the three angles shown?

Part B: What is the measure of angle *a*?

5. Draw lines of symmetry for each letter. Circle the letters that do not have a line of symmetry.

6. Rex claims he can draw a right triangle that has one right angle, one acute angle, and one obtuse angle. Is Rex correct? Explain.

132 Grade 4 · Chapter 14 Geometry

7. Three angles combine to form a straight angle. Which of the following must be true? Select all that apply.

☐ One of the angles must be obtuse.
☐ At least two of the three angles must be acute.
☐ All three angles must be acute.
☐ None of the angles can be obtuse.
☐ All three angles can be acute.
☐ One of the angles can be a right angle.
☐ One of the angles must be a right angle.

8. Angles x and y combine to form a 120° angle. The angle measure of y is twice the angle measure of x. What are the measures of angle x and angle y? Show your work.

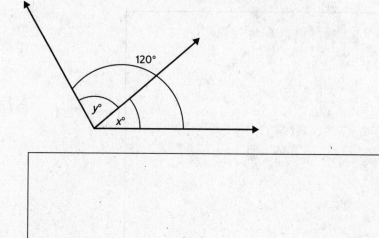

Grade 4 · Chapter 14 Geometry   133

9. A quadrilateral has two obtuse angles that are equal in measure and two acute angles that are equal in measure. Draw the quadrilateral.

10. A quadrilateral has two right angles and one pair of parallel line segments.

   Part A: Draw the quadrilateral and label its right angles. Is the quadrilateral a rectangle or a parallelogram? Explain.

   Part B: Which line segments in the quadrilateral are parallel?

11. Find the measure of each angle in the quadrilateral. What is the sum of all four angles?

12. Warren drew a diagonal line to connect two opposite corners of a rectangle. Then he cut the rectangle in two along that diagonal line. What figures were formed? Which figure is greater in size? Use a diagram to explain your answer.

13. Write the number of lines of symmetry that each figure has in the box.

	an oval
	a person
	a glove
	the letter P
	an 8-sided stop sign
	a parallelogram

14. Will the gray tile completely fill in the shaded space without any gaps or overlaps? If so, how many tiles will it take? Make a model to obtain your answer.

15. Which quadrilaterals can have at least one pair of parallel sides, at least one pair of sides that are equal in length, and at least one right angle?

Grade 4 · Chapter 14 Geometry

16. Dania started with her angle arrow in the position shown. Then she took these four steps:

   Step 1: $\frac{1}{4}$ turn left

   Step 2: $\frac{1}{2}$ turn right

   Step 3: $\frac{1}{4}$ turn right

   Step 4: more than $\frac{1}{2}$ turn left

   In what position does Dania's angle arrow end up? Draw the arrow.

17. Write "parallel," "perpendicular," or "neither" for each pair of streets.

   ONLINE TESTING
   On the actual test, you may drag items into boxes. In this book, you will write the values in the boxes.

	Roy Street and Lake Lane
	Roy Street and 1st Street
	Pine Street and Roy Street
	K Road and 3rd Street
	1st Street and 3rd Street
	Lake Lane and 2nd Street

   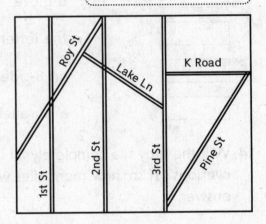

18. Indicate whether each statement is true or false.

True	False	
☐	☐	A half-turn is equal to 180°.
☐	☐	A right triangle has two right angles.
☐	☐	A parallelogram can have a right angle.
☐	☐	A rhombus can be a square, but a square cannot be a rhombus.
☐	☐	A trapezoid always has one pair of sides with equal lengths.

Performance Task

Get to Know a National Park

Anna is doing a school project on Yellowstone National Park. She also plans to visit the park with her family this summer.

Write your answers on another piece of paper. Show all your work to receive full credit.

Part A

The table shows the number of visitors that Yellowstone has had during the first five months of the year. Draw a bar graph below the table that shows the number of visitors for each month rounded to the nearest thousand.

Month	Number of Visitors to Yellowstone
January	26,778
February	28,233
March	18,778
April	31,356

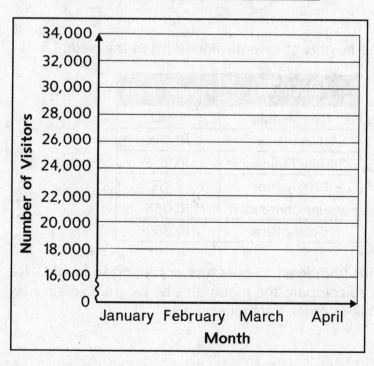

Grade 4 • Chapter 1 Place Value 137

Performance Task (continued)

Part B

In May, between 300,000 and 400,000 people visited Yellowstone. In the total number of visitors for May, the digit 1 has 10 times the value that it has in the total number of visitors in April. In the total number of visitors for May, the tens digit and the thousands digit are both the same even number, and the ones digit is three times the tens digit. The only other digit is a zero, and it is the only zero in the number. How many people visited Yellowstone in May? Explain your reasoning.

Part C

Yellowstone covers 3,468 square miles of land. Write the expanded form and the word form for Yellowstone's land area.

Part D

The table shows the heights of several mountains in the park.

Mountain	Height (feet)
The Needle	9,862
Hoyt Peak	10,344
Saddle Mountain	10,394
Grizzly Peak	9,915
Mount Chittenden	10,088
Castor Peak	10,804

Order the mountains from **least** to **greatest** according to height. Use inequality symbols to compare the mountains by height. Explain how you used place value to order the mountains.

138 Grade 4 · Chapter 1 Place Value

NAME _____   DATE _____   SCORE _____

Performance Task

The Science Club

The science club sponsors field trips and science competitions during the school year. The club conducts fundraising activities to raise money to pay for the events.

Write your answers on another piece of paper. Show all your work to receive full credit.

Part A

Nine students and one adult plan to go on the club's trip to the natural history museum. Student tickets cost $6 each and adult tickets costs $10 each. Lunch at the museum cafe costs $5 per person. Each person will get a lunch from the cafe. The museum also offers a group package for 12 people of any age that includes a museum ticket and lunch for a total of $90. The club wants to spend the least amount of money possible. Should it purchase individual lunches and tickets or the group package? How much less will the club pay by choosing that option? Explain your reasoning.

Part B

Club members will participate in a televised science competition in December. The school has made 5 buses available to transport students to watch the competition in the studio audience. The buses can transport a total of 260 students to the competition. So far, 33 first graders, 119 second graders, 29 third graders, and 38 fourth graders have signed up to ride a bus. What is the greatest number of students that can still sign up to ride a bus to the competition? Explain your reasoning.

Grade 4 · Chapter 2 Add and Subtract Whole Numbers **139**

Performance Task (continued)

Part C

After a fundraising dance, the science club treasurer must deduct the costs of the dance from the money taken in. She writes a subtraction problem in the balance book to calculate the new balance. When the balance book accidentally gets wet, two digits in the problem are erased. The subtraction problem then looks like this:

$$2,082 - 6\square5 = 1,\square47$$

What are the missing digits? Use your knowledge of place value to explain your reasoning.

Part D

A local organization will donate the same amount of money to the club's trip fund each week in January. The club started January with some money already in its fund. Fill out the table with the missing dollar amounts. How much did the science foundation donate each week? Explain your reasoning.

End of Week	Amount in Fund ($)
0	
1	237
2	326
3	415
4	

Part E

The club plans a year-end trip to the Smithsonian museums in Washington, DC. At the end of February, the club had $537 in its account. The trip costs include $3,358 for meals and lodging, $1,478 for transportation, and $600 for entertainment. A donor has offered to pay $2,000 of the costs. Write an equation that can be used to find the remaining amount of money that the club must raise. Use n to represent the remaining amount. Then solve the equation. Explain your reasoning.

Performance Task

Crafty Artists

Students in the Future Artists club decorate and sell t-shirts and mugs. The table shows the supplies that the club ordered during November.

Item	Number Ordered	Cost per Item
plain t-shirt	40	$3
plain mug	24	$2
box of fabric markers	24	$10
box of paint markers	36	$12

Write your answers on another piece of paper. Show all your work to receive full credit.

Part A

The club orders an *additional* 2 cartons of fabric markers. Each carton contains 8 boxes of markers. To find the total cost, Ian first calculated 2 × 8. He then multiplied the result by the cost of each box. Mila also calculated the total. She first found the cost of 8 boxes of markers. She then multiplied the result by the number of cartons. Write a single number statement that shows that Ian and Mila's totals are equal. Explain the property that the number statement demonstrates.

Part B

Write a word statement that compares the cost of a box of paint markers to the cost of a plain t-shirt. Then write a number statement involving multiplication that shows the same comparison.

Grade 4 • Chapter 3 Understanding Multiplication and Division

Performance Task (continued)

Part C

Plain t-shirts come in packages of 4. In order to find *p*, the number of packages of t-shirts in the order, Jada writes the equation $40 \div 4 = p$. Write two *different* equations, one using multiplication, and the other using division, that could also be used to find *p*. Then find the value of *p*.

Part D

The club ordered three times as many boxes of paint markers in November as it did in September. In the diagram, each marker represents one box of markers. Show how you can find the number of boxes of paint markers the club ordered in September. Then state the number of boxes ordered in September.

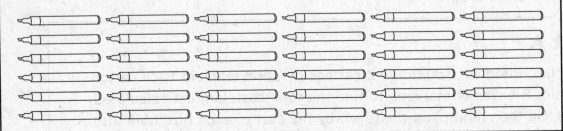

Part E

The club members decorated all of the mugs ordered in November. They arranged the mugs on a table at a craft fair. The arrangement is shown viewed from above. Draw two *different* ways that the mugs can be arranged in rows so that each row has the same number of mugs.

Performance Task

Ocean Camp

This summer, Bryan is spending a month at an ocean camp that is 428 miles from his home. He is studying ocean wildlife and participating in recreational activities.

Write your answers on another piece of paper. Show all your work to receive full credit.

Part A

Bryan's parents drove him to the camp in the family car, and then returned home. They will drive the car the same route to pick Bryan up at the end of the month. Bryan's older sister Lily plans to drive the family car to the camp in the middle of the month to visit him, and then return home. How many total miles will the car be driven to and from the camp this summer? Show your work.

Part B

Lily says that she can usually drive 62 miles per hour on the highway that leads to the camp. Is it reasonable to assume that Lily can make it to the camp in 6 hours if she drives at that rate? Explain your answer using estimation.

Part C

Bryan is studying gray whales, which migrate north and south along the Pacific coast of North America each year. The round is trip approximately 12,000 miles. Explain how you can use multiplication of a one-digit number by a two-digit number to find the approximate number of miles a gray whale migrates along the Pacific coast in 5 years.

Performance Task (continued)

Part D

Bryan volunteers for one full week feeding sea lion pups at the marine sanctuary. There are 4 pups that each eat 13 pounds of fish per day. Complete the area model to show the total number of pounds of fish that Bryan feeds the sea lion pups for one week. Then write the partial products and the total product.

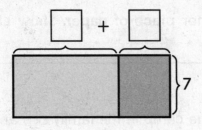

Part E

Bryan is going on an all-day sea kayaking adventure with 15 other campers and two camp leaders. Each person will get a kayak. Each kayak weighs 53 pounds. The table shows the maximum weights that 5 different kayak trailers can haul. The group will use one kayak trailer. Use estimation to figure out which trailers can haul all of the kayaks needed for the adventure. Explain your answer.

Trailer	A	B	C	D	E
Maximum Weight (pounds)	1,250	600	900	1,500	1,050

144 Grade 4 · Chapter 4 Multiplying

Performance Task

Happy Paws

Jen and Quon are helping out Happy Paws animal rescue for their school service project.

Write your answers on another piece of paper. Show all your work to receive full credit.

Part A

Jen and Quon asked a pet food company for food donations. The company agreed to donate 85 pounds of dog food each month for a year. Complete the area model for the total number of pounds of dog food that the company will donate in one year. Then use the model to find the total.

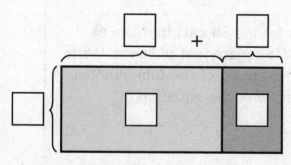

Number of Pounds of Dog Food:

Part B

The company also agreed to donate 95 pounds of cat food to Happy Paws each month for a year. Explain how to use your answer from **Part A** to find the total number of pounds of cat food that will be donated in one year. Then state the total.

Performance Task (continued)

Part C

Happy Paws has 53 dogs in foster homes. The director has asked Jen to pack a box of canned dog food for each foster home. Each box will contain 25 cans. Jen does the multiplication shown to figure out how many cans she will need. Use place value to explain the error that Jen made. Then correct the error and find the total number of cans.

```
    53
  × 25
   265
 + 106
   371
```

Part D

Quon is helping set up an outdoor cat pen for the 38 cats that live at Happy Paws. Each cat should have at least 16 square feet of space. Write an equation that can be used to find s, an estimate of the total number of square feet the pen should enclose. Then solve the equation.

Part E

Is your estimate from **Part D** greater than or less than the actual number of square feet? Explain. In this case, explain whether it better for the estimate to be less than or greater than the actual number.

146 Grade 4 · Chapter 5 Multiply with Two-Digit Numbers

Performance Task

Hiking the Pacific Crest Trail

Rodrigo is a college student. He plans to section hike the length of the Pacific Crest Trail with 5 friends. The trail runs from Mexico to Canada through California, Washington, and Oregon.

Write your answers on another piece of paper. Show all your work to receive full credit.

Part A

The friends will hike through sections of the trail during 3 summers, over a total of 9 months. They want to cover about the same distance each month. Research the length of the trail in miles. Use compatible numbers to estimate the number of miles they will hike each month. State whether your estimate is greater than or less than the actual number of miles they would hike each month.

Part B

Now estimate the distance the friends will hike each month by rounding the numbers to the largest possible place value. Is this distance less than, equal to, or greater than your estimate from **Part A**? Why do you think this is so?

Grade 4 · Chapter 6 Divide by a One-Digit Number

Performance Task (continued)

Part C

Along their route, the hikers plan to climb to the top of Mt. Hood in Oregon. They will allow 10 hours to reach the summit. Research the height of Mt. Hood in feet. Use the closest multiple of 10 for the height. To reach the summit, how many feet in elevation must they gain each hour from their base camp of 8,500 feet?

Part D

The hikers will carry backpacks with their own personal gear as well as group items whose weight will be equally shared among them. The shared items weigh a total of 78 pounds. Write an equation that can be used to find w, the weight of the shared items carried by each hiker. Then solve the equation.

Part E

Rodrigo and his friends plan to take one group photo every time they go through a major mountain pass along the trail. Research the number of major mountain passes along the Pacific Crest Trail. Is it possible for each of them to take the same number of such photos? Use your knowledge of remainders to explain why or why not.

Performance Task

Designing Store Displays

Cearra is helping her parents at the local market that they own.

Write your answers on another piece of paper. Show all your work to receive full credit.

Part A

Cearra makes a display with soup cans. The first three rows of the display are shown. Extend the pattern by drawing two more rows of cans.

Row 3
Row 2
Row 1

Part B

Write an equation that is a rule for finding c, the number of cans in row number r of the display. Explain how you figured out the equation.

Grade 4 • Chapter 7 Patterns and Sequences

Performance Task (continued)

Part C

Use your equation to find the number of cans in Row 7. Show your work.

Part D

How many total rows will there be when Cearra is finished with the soup can display? Explain how you know.

Part E

Cearra makes another display with piles of lemons that form a pattern. The first three piles are shown. Complete the table. Then state a rule for the pattern.

Pile Number	1	2	3	4	5
Number of Lemons	3				

Rule:

Performance Task

The River Bank Market Center

River Bank Market is an urban shopping and entertainment center that is still in its planning stage. River Bank has sold parts of its 1000 square meter space to 6 different client groups. Nerd Nation Fitness has purchased $\frac{7}{30}$ of the space for its center. G-Street Apps has an area equal to $\frac{1}{6}$ of the total space. Bike Geek Cycle Shop has $\frac{1}{30}$ of the center's total space for a shop. Dora's Unbelievably Great Cupcakes has $\frac{1}{10}$ of the space. The Green Door Theater has $\frac{5}{12}$ of the total space. Noodlehead's has $\frac{1}{20}$ of the space.

Part A

The first problem that the center is looking to solve is to determine how much renters should pay. Before they can determine an actual price, the planners must be able to rank the renters in order of the amount of space they are renting. List each renter client from greatest to least with respect to the amount of space that they are renting.

Part B

The River Bank planners need to draw up a floor plan of the building to see how each space lines up. So far, the architects have come up with the floor plan shown below that divides the rectangular space into equal-sized parts that will be divided among the clients. Color in the squares to show a way in which the space for each client can be laid out.

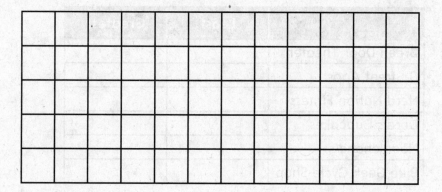

Performance Task (continued)

Part C

The clients want to see another variation of the floor plan you made. Draw an additional floor plan and color it in to show each client's space.

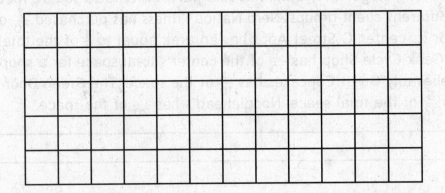

Part D

Having gone over the previous plans, the clients now want the architect to come up with as many rectangular plans for the entire complex as possible. How many different rectangles can you draw? Make a sketch of each rectangle showing the individual squares as you drew above. Each rectangle must be at least 3 squares in width.

Part E

The builders have decided to charge the renters between $850 and $1,000 per square unit as shown in the diagram in **Part B**. Determine the least and greatest amount of rent each client will pay.

	Least	Greatest
Green Door Theater		
G-Street Apps		
Nerd Nation Fitness		
Dora's Cupcakes		
Noodlehead's		
Bike Geek Cycle Shop		

Performance Task

Ask Omar

Omar is the author of the website, Ask Omar the Baker, which answers questions about baking. Here is today's question:

Hey Omar, My recipe for biscuits calls for $5\frac{1}{3}$ pounds of flour. But I have only one measuring cup and it measures $\frac{2}{3}$ of a pound of flour at a time. How can I figure out how to measure $5\frac{1}{3}$ pounds of flour using my measuring cup? Signed, Biscuit Betty

Write your answers on another piece of paper. Show all your work to receive full credit.

Part A

Use a model to show how Betty can measure the $5\frac{1}{3}$ pounds of flour using her $\frac{2}{3}$ lb measuring cup. Then write a response from Omar to Betty describing how she can solve the problem.

Performance Task (continued)

Part B

Gourmet Gabby writes that her recipe uses $2\frac{2}{3}$ fewer pounds of flour than Biscuit Betty's recipe. Gabby's measuring cup holds $\frac{1}{6}$ of a pound. How many measuring cups full of flour does Gabby need?

Part C

Ahmed writes that he needs to measure 8 pounds of flour. To make an exact measurement with no flour left over, which measuring cup should Ahmed use, one that holds $\frac{3}{5}$ pound or $\frac{4}{5}$ pound? Explain.

Part D

Toni Mahoney is using Biscuit Betty's recipe that requires $5\frac{1}{3}$ pounds of flour to make biscuits for her brunch. Suddenly, Toni learns that 4 times as many people are coming to her brunch than originally planned. How much flour will Toni need to make her biscuits? Explain how you got your answer.

Performance Task

Wendy's Weather Station

Wendy's Weather Website collects data from weather volunteers. The map shows rain totals in inches for Wednesday. Wendy would like to post the totals, but she sees that some volunteers gave their total in fraction form and some gave totals in decimal form.

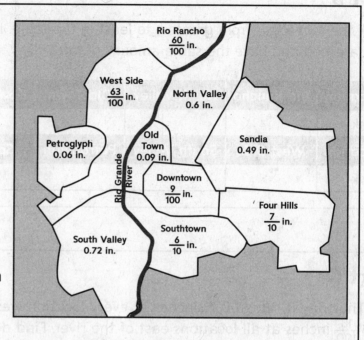

Write your answers on another piece of paper. Show all your work to receive full credit.

Part A

Fill in the table to post rain totals for Wednesday.

Location	Rio Rancho	North Valley	Old Town	Downtown	Southtown
Rain (in) fraction					
Rain (in) decimal					

Location	West Side	South Valley	Petroglyph	Sandia	Four Hills
Rain (in) fraction					
Rain (in) decimal					

Grade 4 · Chapter 10 Fractions and Decimals

Performance Task (continued)

Part B

Rank the rain totals from greatest to least in the table. Indicate locations that are tied and have the same amount of rainfall.

Location	Rio Rancho	North Valley	Old Town	Downtown	Southtown
Rank					
Location	West Side	South Valley	Petroglyph	Sandia	Four Hills
Rank					

Part C

On Thursday it rained 0.15 inches at every location west of the river and $\frac{18}{100}$ inches at all locations east of the river. Find decimal totals for each location for Wednesday and Thursday combined.

Location	Rio Rancho	North Valley	Old Town	Downtown	Southtown
Rain (in) Wednesday					
Rain (in) Thursday					
Total (in) Weds/Thurs					
Location	West Side	South Valley	Petroglyph	Sandia	Four Hills
Rain (in) Wednesday					
Rain (in) Thursday					
Total (in) Weds/Thurs					

Performance Task

Manny the Mover

On the side of Manny the Mover's truck, it says he can move "anything, anywhere, anytime." Manny's current order features boxes that measure 3 feet by 2 feet by 2 feet. The floor of Manny's truck bed measures 8 yards in length and 96 inches in width.

Write your answers on another piece of paper. Show all your work to receive full credit.

Part A

Draw and label a floor plan model of Manny's truck bed. Model tiles to represent boxes are shown. Use a grid pattern to show each square unit in your diagram.

[Diagram: a 3 ft by 2 ft rectangle tile, and a 2 ft by 2 ft square tile]

Part B

Show 3 different plans labeled Plan 1, Plan 2, and Plan 3 for packing 1 layer of boxes on the truck. Assume that you can squeeze boxes into tight spaces. For example, you can squeeze a 2-foot wide box into a 2-foot space.

Grade 4 • Chapter 11 Customary Measurement

Performance Task (continued)

Part C

Which plan from **Part B** allows you to pack the greatest number of boxes? Explain.

Part D

The height of Manny's cargo space is 2 yards and 12 inches. If Manny wants to choose the plan that can fit the most boxes with the smallest amount of gaps, which plan should choose? Stack the boxes in layers of equal height. How many boxes will each plan allow? Will there be any gaps? Explain.

Part E

Manny has just discovered that his special tall truck with a 108-inch ceiling is available. If he uses the tall truck, does this change the number of boxes that he can pack? Explain.

Performance Task

Frog House

A zoo is designing a new Frog House that will house leopard frogs, bullfrogs, spring peepers, toads, and poison dart frogs.

Write your answers on another piece of paper. Show all your work to receive full credit.

Part A

Here are the dimensions for each frog's rectangular habitat.
- Bullfrogs: 400 cm by 6000 mm
- Leopard frogs: Same width as bullfrog, but 1 meter shorter in length
- Toads: 250 cm shorter in both length and width than leopard frogs
- Spring peepers: 1,500 mm greater in length than toads and 50 cm less in width than toads
- Poison dart frogs: Same length as spring peepers, but 3 m greater in width than peepers

Find the dimensions of each frog's habitat in meters.

Performance Task (continued)

Part B

Draw and label a rectangular floor plan for the Frog House that measures 0.02 kilometers by 10 meters. Label your floor plan in meters.

Part C

Here are the special requirements for the habitats. Draw in each habitat on your floor plan.
- All habitats: Must have at least a 2 m distance from any other habitat
- All habitats: Must have at least a 100 cm distance from any outer wall of building
- Bullfrogs: Must be 3 m from any outer wall of the building
- Poison dart frogs: Must have at least a 1,000 cm distance from spring peepers
- Toads: Must have at least a 1,000 cm distance from leopard frogs

Performance Task

Square Deal

Perimeter and area are two ways to measure the size of a figure. In this activity, you will explore patterns of perimeter and area.

Write your answers on another piece of paper. Show all your work to receive full credit.

Part A

Explore a pattern of squares. Create models to show how squares increase in perimeter and area as their side length increases. The first three are done for you. Draw the next seven.

Part B

Now show the numerical patterns of the series. Refer to the models you drew in **Part A** to fill in the table. In words, describe the pattern for both perimeter and area. Which pattern increases in value faster, perimeter or area? Explain.

Length (ft)	Width (ft)	Perimeter, P (ft)	Area, A (sq ft)
1	1	4	
2	2		
3	3		
4	4		
5	5		
6	6		
7	7		
8	8		
9	9		
10	10		

Grade 4 • Chapter 13 Perimeter and Area

Performance Task (continued)

Part C

Now, analyze the patterns for perimeter and area. Find the differences in perimeter and area as the squares increase in side length. Fill in the table. Then describe how each pattern changes.

Length (ft)	Width (ft)	Perimeter, P (ft)	Perimeter difference	Area, A (sq ft)	Area difference
1	1	4	—	1	—
2	2	8	4	4	3
3	3	12	4	9	5
4	4	16	4	16	7
5	5	20	4	25	9
6	6	24	4	36	11
7	7	28	4	49	13
8	8	32	4	64	15
9	9	36	4	81	17
10	10	40	4	100	19

Part D

Explore a different pattern in which squares double in both length and width. Fill in the doubling table. Describe the patterns that you see.

Length (ft)	Width (ft)	Perimeter, P (ft)	Area, A (sq ft)
1	1	4	1
2	2	8	4
4	4	16	16
8	8	32	64
16	16	64	256
32	32	128	1,024

Performance Task

Triangle Truth

Do all triangles have characteristics in common? In this activity, you will make some discoveries by drawing and measuring triangles and other figures.

Write your answers on another piece of paper. Show all your work to receive full credit.

Part A

Draw triangles 1-6 with the following characteristics:

1. An acute triangle in which all three sides have the same length
2. An acute triangle in which all three sides have different lengths
3. An obtuse triangle in which two sides have the same length
4. An obtuse triangle in which all three sides have different lengths
5. An right triangle in which two sides have the same length
6. An right triangle in which all three sides have different lengths

Part B

Use a protractor to measure the angles of each triangle. Be precise. Record your measurements in the table.

	Triangle type	Equal Sides	Angle 1	Angle 2	Angle 3
1	acute	3 equal sides			
2	acute	no equal sides			
3	obtuse	2 equal sides			
4	obtuse	no equal sides			
5	right	2 equal sides			
6	right	no equal sides			

Performance Task (continued)

Part C

Find the sum the angle measures for each triangle. Record your calculations. What pattern do you see? Explain.

Triangle type	Equal Sides	Angle 1	Angle 2	Angle 3	Sum
1	acute	3 equal sides				
2	acute	no equal sides				
3	obtuse	2 equal sides				
4	obtuse	no equal sides				
5	right	2 equal sides				
6	right	no equal sides				

Part D

What hypothesis do you make from your data from **Part C**? How could you find more evidence to support your hypothesis?

Benchmark Test 1

1. The Garcias are buying a car priced between $10,000 and $20,000. The car's price is a multiple of 100. The hundreds digit is 5 times the thousands digit.

   Part A: Write the price of the car in word form.

   [   ]

   Part B: Fill in the blanks to complete the expansion of the car's price.

   ☐ × [   ] + ☐ × 1,000 + ☐ × 100 + ☐ × 10 + ☐ × 1

2. The table shows the numbers of 4 different types of CDs sold at a chain of music stores. Select > or < to compare the indicated numbers.

Type of CD	Number Sold
Pop	20,756
Country	19,984
Rock	21,001
Hip Hop	20,689

   > <
   ☐ ☐ the number of country CDs _____ the number of hip hop CDs
   ☐ ☐ the number of pop CDs _____ 20,000
   ☐ ☐ the number of rock CDs _____ the number of hip hop CDs
   ☐ ☐ the number of country CDs _____ 20,000
   ☐ ☐ the number of hip hop CDs _____ the number of pop CDs
   ☐ ☐ the number of rock CDs _____ the number of pop CDs

3. Rashan has $23,456 in his college savings account. Joannie has $14,365 in her college savings account. Sort the digits below according to their value in each student's account. Some digits may not be used.

   | 1 | 2 | 3 | 4 | 5 | 6 |

Value of Digits	
Worth 10 times more in Rashan's account than in Joannie's	Worth 10 times more in Joannie's account than in Rashan's

Grade 4 • Benchmark Test 1   165

4. The digit in the hundred thousands place is 4 times the digit in the ones place. The digit in the tens place is 3 times the digit in the ones place. The digit in the thousands place has 10 times **the value** of the digit in the hundreds place. The digit in the ten thousands place has 10 times **the value** of the digit in the thousands place. Fill in the missing digits to make the inequality statement true.

☐☐☐,3☐☐ > 500,000

5. Andrea agreed to sell 100 raffle tickets for a band fundraiser. The table shows the number of tickets she sold during each of the first three weeks of the month. She estimates that she has about 30 more to sell. Use the numbers to write an equation that shows how Andrea could have come up with her estimate. Some numbers may not be used.

Week	Number of Tickets Sold
1	21
2	8
3	37

0   10   20   30   40   50   70   80   100

☐ − (☐ + ☐ + ☐) = ☐

6. Tariq is saving money for a trip to an amusement park. He will need $50 for the ticket and $20 for food. Tariq saves the $42 he earned helping at his uncle's business. His sister gives him a coupon for $10 off admission to the park. Select all expressions that are equal to the remaining amount, in dollars, that Tariq must save.

☐ 50 + 20 − (42 + 10)   ☐ (50 + 20) − (42 + 10)

☐ (50 + 20) − 42 + 10   ☐ (50 + 20) − 42 − 10

☐ 50 + 20 − 42 − 10   ☐ (50 + 20) − (42 − 10)

7. Complete the multiplication problem by writing in the missing digits.

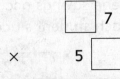

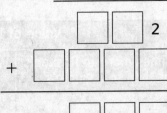

8. Martina and Barry rented a boat. They filled it with 18 gallons of gas. They used 6 gallons of gas to get to an island and 3 gallons waterskiing and sightseeing.

   Part A: Write an equation that can be used to find *g*, the number of gallons of gas that are left. Use the numbers and variable shown to fill in the boxes.

   | 3 | 6 | 18 | *g* |

   ☐ − (☐ + ☐) = ☐

   Part B: Solve the equation.

9. The school dance took in $1,100. Expenses totaled $304. The dance advisor wrote a subtraction problem to figure out the amount of profit from the dance, after expenses. Select whether each statement describes a step that can be used to find the answer to the advisor's subtraction problem.

Yes	No	
☐	☐	Subtract 0 tens from 0 tens, which leaves 0 tens.
☐	☐	Subtract 0 thousands from 1 thousand, which leaves 1 thousand.
☐	☐	Use one hundred as 10 tens.
☐	☐	Use one ten as 10 ones, which leaves 0 tens and 10 ones.
☐	☐	Subtract 4 ones from 10 tens.
☐	☐	Use one ten as 10 ones, which leaves 9 tens and 10 ones.

10. Ms. Albright has 42 packages of paper. Each package has 50 sheets of paper. Mrs. Albright wants to find the total number of sheets of paper she has. Which is the best number sentence for her to use as a first step?

    A. 42 × 5 = 210   C. 42 + 50 = 92

    B. 42 + 5 = 47   D. 42 × 15 = 630

Grade 4 • Benchmark Test 1 **167**

11. A gift shop owner has 21 candles. She puts the same number in each of 7 gift baskets.

   Part A: Write a division statement whose answer is the number of candles in each basket. Include the answer in your statement.

   Part B: Represent the division statement on a number line.

   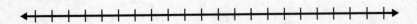

12. A farmer arranged some pumpkins as shown. Show 2 more arrangements that use the same total number of pumpkins. Within an arrangement, each row should have the same number of pumpkins.

   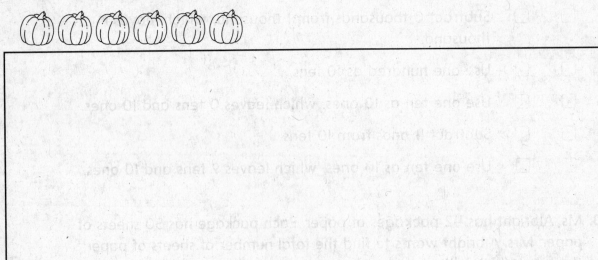

168 Grade 4 • Benchmark Test 1

13. Alice has 8 dimes. Noah has 24 dimes. The equation 24 = 8 × 3 can be used to describe the relationship between the number of dimes that Alice and Noah have. Which statement below also represents this relationship?

   A. Noah has 3 more dimes than Alice.

   B. Noah has 3 times as many dimes as Alice.

   C. Alice has 3 more dimes than Noah.

   D. Alice has 3 times as many dimes as Noah.

14. Hitomi, Julius, and five of their friends each have the same number of crayons. The total number of crayons they have is shown. Circle groups of crayons to model the number the each student has.

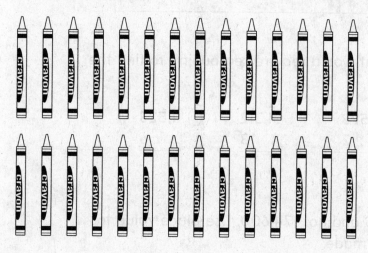

15. Look at the area model. Complete the number sentence so that it shows the partial products and final product shown by the model. Fill in the missing numbers.

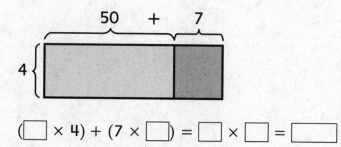

(☐ × 4) + (7 × ☐) = ☐ × ☐ = ☐

Grade 4 • Benchmark Test 1 **169**

16. A catering company will earn the same amount on each of the 3 days of an event. To figure out about how much it would earn, the company estimated the product ___ 82 × 3 as 2,400 by first rounding to the greatest place value possible.

    Part A: What is the value of the missing digit?

    Part B: Is the estimate greater than or less than the actual number? Explain.

17. Select **all** of the problems that would require regrouping to find the product.

    ☐ 322 × 5   ☐ 241 × 4   ☐ 3,523 × 2   ☐ 31,312 × 3   ☐ 4,523 × 6

18. Desiree multiplied 3,492 by 5 and got 174,600. Use place value to explain the mistake that she made.

19. A building has *f* floors. Each floor has 34 offices. The number *f* is an odd 2-digit number. Indicate whether each statement is true or false.

True False

☐ ☐ The total number of offices in the building could be a multiple of 10.

☐ ☐ The number of floors must be a multiple of 5.

☐ ☐ The total number of offices in the building is greater than 350.

☐ ☐ The total number of offices in the building is divisible by *f*.

☐ ☐ The number equal to 34 × *f* could be greater than 3,400.

20. A company ordered 23 sandwich platters and 32 salads for a meeting. Sandwich platters cost $13 each, and salads cost $7 each. The company paid a $6 delivery fee.

   Part A: Use some of the numbers and symbols shown to complete the equation so that it shows the total amount, *n*, that the company paid for the order.

   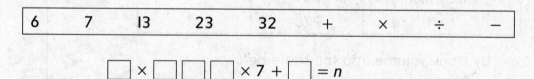

   Part B: Solve the equation. Show your work and state the answer.

Grade 4 • Benchmark Test 1   **171**

Performance Task

Giant Trees

A biologist is studying giant trees for a conservation project. She needs help finishing her data collection and comparing different values.

Write your answers on another piece of paper. Show all your work to receive full credit.

Part A

Look up the trunk volume of the following giant sequoia trees. Complete the table. Use values for trunk volume that do not include limbs and loss from burns.

Tree Name	Height (feet)	Trunk Volume (cubic feet)
Boole	269	
General Grant	267	
General Sherman	248	
Grizzly Giant	209	
Hart	278	

Sort the trees by trunk volume into the table below.

Volume (in Cubic Feet) of the World's Largest Sequoias			
< 40,000	> 40,000 < 45,000	> 45,000 < 50,000	> 50,000

Part B

The biologist finds a young giant sequoia tree. The tree in the table with the largest volume has a volume that is 5 times the volume of the younger tree. Write an equation that can be used to find y, the volume of the younger tree. Then solve the equation.

Benchmark Test 2

1. A bookshop owner must place 72 books equally on 6 different shelves. The model below shows how he did this.

   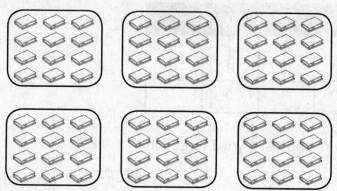

   Part A: Complete the division sentence shown by the model.

   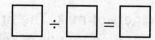

   Part B: What does the answer to the division sentence mean in terms of the problem situation?

2. Look at the number list.

   Part A: Extend the pattern.

   27, 30, 21, 24, 15, 18, _____, _____

   Part B: Complete the statement.

   The pattern is _____

3. Jocelyn has $\frac{4}{6}$ cup of lemon juice. She adds an additional $\frac{3}{9}$ cup to it.

   Part A: Write each fraction in lowest terms.

   Part B: How much lemon juice does she have in all?

Grade 4 • Benchmark Test 2   **173**

4. An electrician has 5 lengths of wire. The table shows the lengths in yards.

   Part A: Rank the wires from least (1) to greatest (5) on the table.

Wire	A	B	C	D	E
Length (yd)	$\frac{1}{4}$	$\frac{1}{16}$	$\frac{5}{16}$	$\frac{1}{8}$	$\frac{3}{8}$
Rank					

   Part B: Where would a wire that is $\frac{5}{8}$ yard long rank?

5. Mrs. Johansen has two gallon-sized containers of milk. The first one has $\frac{5}{6}$ gallon in it. When she pours the second container of milk into the first, she gets exactly 1 gallon. How many gallons were in the second container?

6. Mark "Yes" for **each** division statement that has a remainder and "No" for **each** that does not.

Yes	No	
☐	☐	120 ÷ 10
☐	☐	66 ÷ 5
☐	☐	131 ÷ 4
☐	☐	63 ÷ 7
☐	☐	132 ÷ 11

7. Giovanni conducted a survey of how far people drive to work. The results are shown in the table.

Town	Drive to Work More than 10 miles	Drive to Work Less than 10 miles
Springfield	15	30
Jacksontown	22	18
Blakesville	10	40

Part A: The ratio of drivers who drive more than 10 miles to those who drive less than 10 miles for the town of Clarksville is the same as for the town of Blakesville. If 20 people surveyed in Clarksville drive more than 10 miles to work, how many people in the Clarksville survey drive less than 10 miles to work?

Part B: Which town had more than half of their drivers driving more than 10 miles to work?

8. Janice listed the first 6 multiples of 5 on the board. Select **all** correct statements.

☐ The ones digit in all of the numbers is 5.

☐ After 5, the tens digit follows the pattern 1, 1, 2, 2, 3, 3, ...

☐ The sum of the digits in each number in the list is divisible by 5.

☐ Every other number ends in 0.

☐ Each number in the list is divisible by 5.

9. Franco bought a notepad for $0.85. Circle the smallest number of coins that Franco can use to pay for the notepad.

10. A scientist weighed a lizard specimen and found it to weigh 0.5 ounces. Circle all the numbers that have the same value as 0.5.

    0.50 $\frac{5}{10}$ $\frac{5}{100}$ $\frac{50}{100}$ $\frac{1}{2}$

11. A sports shop owner stocks several different kinds of jerseys. The table below shows how many of each type he has in stock. He is organizing the jerseys onto racks that will contain a variety.

Type of Jersey	Number of Jerseys
Football	47
Soccer	38
Baseball	41
Volleyball	32

Part A: On one set of racks he wants to hang the football and soccer jerseys. Each rack can hold 9 shirts. How many racks does the owner need? What is the significance of the remainder?

Part B: The owner has eight racks to dedicate to the baseball and volleyball jerseys. Can he fit them all? If not, how many will be left over?

12. A stamp collector buys 7 stamps during his first year of collecting stamps. After the first year, he buys 6 more stamps than he did the previous year.

    Part A: How many stamps did he purchase in the 5th year?

    Part B: Find the total number of stamps that he purchases in the first 5 years. Show your work.

13. Write an addition sentence for each model. Then find the sum.

    Part A:

    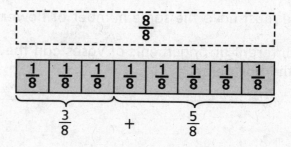

    _____

    Part B:

    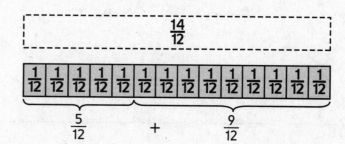

    _____

    Part C:

    _____

    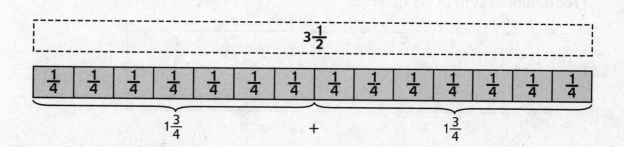

14. A painter has two containers of paint. The first container has 0.67 gallons in it. The second has 0.3 gallons in it. Shade the models below and find the total amount of paint the painter has.

15. A florist is trying to split 36 flowers in vases with the following criteria.

    1. Each vase must have more than 2 flowers.

    2. Each vase must have the same number of flowers.

    How many different arrangements of vases can the florist form? List them all in the box below.

16. Theo has raised $819 for charity. He wants to donate the same amount of money to each of nine charities. He uses the division statement $810 ÷ 9 to estimate that he will give about $90 to each charity. Is this a reasonable estimate? Explain.

178 Grade 4 • Benchmark Test 2

17. Look at the table.

    Which equation describes the pattern?

Input (r)	10	12	15	20
Output (v)	15	17	20	25

    A. $r \times 5 = v$

    B. $v \times 5 = r$

    C. $r + 5 = v$

    D. $v + 5 = r$

18. Determine whether each statement is true or false.

    True False

    ☐ ☐ $\frac{3}{9}$ is equivalent to $\frac{1}{3}$.

    ☐ ☐ $\frac{3}{6}$ is the simplest form of $\frac{12}{24}$.

    ☐ ☐ $1\frac{8}{12}$ is equivalent to $\frac{5}{3}$.

    ☐ ☐ $\frac{7}{16}$ is the simplest form of $\frac{14}{32}$.

19. Enrique rides his bike $\frac{3}{4}$ mile to school every day and the same $\frac{3}{4}$ mile home. If he has school five days each week, circle all of the expressions that will find the total number of miles Enrique rides every week.

    $5 \times \frac{3}{4}$ $5 \times \frac{6}{4}$ $5 \times \frac{3}{2}$

    $10 \times \frac{3}{4}$ $10 \times \frac{3}{2}$ $\frac{6}{4} + \frac{6}{4} + \frac{6}{4} + \frac{6}{4} + \frac{6}{4}$

20. A grocer unpacks boxes of bananas that weigh 87.24 pounds altogether. He started working on packaging the bananas at 9:30 A.M. When he finished filling the bin with all the bananas that would fit, there were 6.21 pounds of bananas left over. With this information, which of the following can you solve? Select all that apply.

    ☐ How many pounds of bananas fit in the bin?

    ☐ When did the grocer finish his work?

    ☐ How long did it take the grocer to put out the bananas?

    ☐ How many boxes of bananas did he unpack?

Performance Task

Selling Tomatoes

A farmer is selling his tomatoes at a local farmer's market. He packages them into bushels and pecks, both of which are measurements of volume. Research how many pecks are in a bushel.

Write your answers on another piece of paper. Show all your work to receive full credit.

Part A

How many pecks are in a bushel? Use this to figure out what fraction of a bushel a peck is.

Part B

The farmer brings 33 pecks. Fill in the number sentence to figure out how many bushels this is. Express the answer as a mixed number.

_____ × _____ = _____ bushels

Part C

If the farmer sells a bushel of tomatoes for $24, for how much should he sell a peck? Show your work.

Part D

The farmer sold 21 pecks in the morning. He sold $4\frac{3}{4}$ bushels in the afternoon. Did he sell more in the morning or in the afternoon? Explain.

180 Grade 4 • Benchmark Test 2

Benchmark Test 3

1. Order the numbers from *greatest to least*.

    654,879   _____

    654,978   _____

    645,879   _____

    645,789   _____

2. Circle the statement that shows *four hundred sixty-four thousand, five hundred four* in expanded form.

    A. 400,000 + 40,000 + 6,000 + 500 + 40

    B. 464,504

    C. 446,450

    D. 400,000 + 60,000 + 4,000 + 500 + 4

3. List all of the factors of 24.

    _____

4. *Part A:* Find the unknown.

   350 + (25 + _____) = (350 + 25) + 19

   Part B: Circle the addition property that is modeled.

   A. Associative Property of Addition

   B. Commutative Property of Addition

   C. Identity Property of Addition

   D. Parentheses Property of Addition

5. Subtract. Check using estimation.

   ```
     608,256
   − 165,704
   ─────────
   ```

6. Use multiplication to complete the number sentence.

   3 times as many

   _____ × _____ = _____

7. List the first five multiples of 11. Explain how you know what the first five multiples are?

   _____

   _____

   _____

   _____

8. Use the clues to complete the place-value chart.

Thousands			Ones		
hundreds	tens	ones	hundreds	tens	ones

The 1 has a value of 1 × 1,000.

The 2 has a value of 2 × 100.

The 4 is in the ten thousands place.

The 6 is in the ones place.

The 7 has a value of 7 × 100,000.

The 9 is in the tens place.

9. Vincenzo joined a basketball league. It costs $23 a week. If the league lasts eight weeks, how much does Vincenzo pay? Show your work.

_____

10. Circle the correct answer.

    2,272 × 4 =

    A. 1,568

    B. 9,088

    C. 8,888

    D. 8,088

Grade 4 • Benchmark Test 3   **183**

11. *Part A:* Estimate.

$71   ⟶
× 52 ⟶   ×_____

Part B: Solve the problem. Is the actual answer *greater than* or *less than* the estimate? Explain.

12. There are 126 pencils in a box. Each classroom in the school gets a box. There are 40 classrooms. How many pencils are there?

Part A: Circle all the correct choices that show a correct way to solve the problem.

A. 126 × 40 =

B.   126
   × 40
   _____

C.   40
  × 126
   _____

D. 40 × 126 =

Part B: Solve the problem.

_____

184   Grade 4 • Benchmark Test 3

13. Divide. Use multiplication and addition to check.

    6)4,381

14. Divide. Use the Distributive Property and the area model to show your work.

    425 ÷ 5 =

    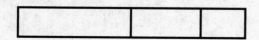

    ____ + ____ + ____ = ____

    425 ÷ 5 = ____

15. Extend the pattern.

    ☐I ☐☐L☐☐☐U

16. Look at the table below.

    Part A: Complete the table.

Input (a)	2	5	8	11	14
Output (z)	14	17	20		

    Part B: Circle the equation that describes the pattern.

    A. $a + 12 = z$

    B. $z + 12 = a$

    C. $a \times 7 = z$

    D. $a + 3 = z$

17. Circle all of the fractions that make the equation true when substituted for x.

    $$\frac{5}{10} = x$$

    A. $\frac{1}{2}$

    B. $\frac{3}{6}$

    C. $\frac{4}{10}$

    D. $\frac{5}{8}$

18. Use the rectangle below to answer the following questions.

   Part A: Shade $\frac{3}{5}$ of the rectangle below.

   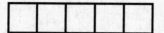

   Part B: Write an equivalent fraction.

19. Tai and her four friends cut some honeydew melons into thirds.

   Part A: If each girl takes $\frac{1}{3}$ of a melon, how much melon would they have taken together? Explain by writing an equation.

   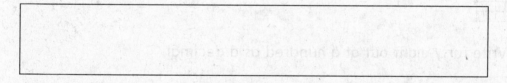

   Part B: If the girls cut 3 melons, how many thirds were left over after they each took their share? Explain by writing an equation.

   Part C: Place a point on the number line for the fraction that represents the total amount of melon taken.

Grade 4 • Benchmark Test 3 **187**

20. Circle all of the fractions that are correct.

    $\frac{5}{8} + \frac{7}{8} =$

    A. $1\frac{1}{4}$ C. $1\frac{4}{8}$

    B. $\frac{3}{2}$ D. $\frac{6}{4}$

21. Emerson has $1\frac{5}{6}$ oranges. She eats $\frac{7}{6}$ oranges. Does she have $\frac{2}{3}$ orange left? Circle yes or no.

    Yes   No

22. Shade the grid model to show forty-eight out of a hundred.

    Part A: Shade the model.

    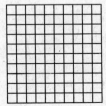

    Part B: Write forty-eight out of a hundred as a decimal.

    Part C: Write forty-eight out of a hundred as a fraction in simplest form.

23. Landon rode 0.87 miles on his scooter. Piper rode 0.78 miles on her bike. Who traveled farther? Explain.

24. Jaylen spends four hours hiking. How many minutes does he hike? Write an equation.

25. Complete the conversion table.

Pints (pt)	Gallons (gal)	(pt, gal)
	2	
	4	
	6	
	8	

26. Complete the conversion table.

Centimeters (cm)	Meters (m)	(cm, m)
100		
400		
600		
800		

27. Choose the best estimate for the length of a surfboard.

   A. 3 meters

   B. 6 meters

   C. 3 kilometers

   D. 6 centimeters

28. Angel is building a sandbox. It will be 3 feet wide and 4 feet long.

   Part A: What is the perimeter of the sandbox? Explain.

   Part B: What is the area of the sandbox? Explain.

Grade 4 • Benchmark Test 3

29. Rian draws an angle with a measure greater than 90° and less than 180°. What type of angle did she draw?

   A. acute

   B. obtuse

   C. perpendicular

   D. right

30. Draw the lines of symmetry in the figure.

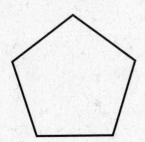

31. Cydney's school is on Main Street, which runs between 4th Avenue and 5th Avenue. Both 4th Avenue and 5th Avenue intersect Main Street at right angles. How would you describe the lines represented by 4th Avenue and 5th Avenue?

   A. Intersecting lines

   B. Perpendicular lines

   C. Obtuse lines

   D. Parallel lines

Performance Task

Organizing a Collection

Ethan and Hali volunteered to help organize the school's collection of rocks and minerals.

Write your answers on another piece of paper. Show all your work to receive full credit.

Part A

Ethan and Hali will use a line plot to record the measurement of each rock's length in inches. Use the lengths in the frequency chart to make a line plot.

Rock and Mineral Length				
2	1	$1\frac{1}{4}$	$1\frac{3}{4}$	$2\frac{1}{4}$
$\frac{2}{4}$	$\frac{3}{4}$	2	2	$2\frac{1}{2}$
$2\frac{1}{4}$	$\frac{1}{2}$	$\frac{3}{4}$	1	$1\frac{3}{4}$

Part B

Look at the line plot. How long is the longest rock? The shortest rock? Which measurement of rock length is plotted the most?

Grade 4 • Benchmark Test 3

Performance Task (continued)

Part C

What is the difference in size between the longest rock and the shortest rock? Explain.

Part D

If a rock is 0.42 inches long, how would you write that as a fraction? Show the number on a number line.

Part E

If one rock is 0.42 inches long, and another rock is 0.29 inches long, which is longer? Show your work as a comparison using >, <, or =. Use a number line to show how you found the answer.

Benchmark Test 4

1. Use the clues to complete the place-value chart.

Thousands			Ones		
hundreds	tens	ones	hundreds	tens	ones

The 1 has a value of 1 × 100.

The 3 has a value of 3 × 1.

The 5 is in the thousands place.

The 7 is in the tens place.

The 8 has a value of 8 × 100,000.

The 9 is in the ten thousands place.

2. Angelica joined a golf league. It costs $37 per week. If the league lasts six weeks, how much does Angelica pay? Show your work.

3. Circle the correct answer.

    1,598 × 5 =

    A. 7,990

    B. 3,950

    C. 6,990

    D. 7,450

Grade 4 • Benchmark Test 4 193

4. 572 + (84 + _____) = (572 + 84) + 65

   Part A: Find the unknown.

   _____

   Part B: Circle the addition property that is modeled.

   A. Associative Property of Addition
   B. Commutative Property of Addition
   C. Identity Property of Addition
   D. Parentheses Property of Addition

5. Subtract. Check using estimation.

   984,065
   − 654,876
   ─────────

6. Use multiplication to complete the number sentence.

   _____ × _____ = _____

7. List the first six multiples of 12. Explain how you know what the first six multiples are.

8. List all the factors of 48.

   [   ]

9. Order the numbers from *greatest to least*.

   835,468   _____

   853,846   _____

   835,846   _____

   853,864   _____

10. Circle the statement that shows *six hundred five thousand, two hundred eleven* in expanded form.

    A. 650,211

    B. 605,211

    C. 600,000 + 50,000 + 200 + 10 + 1

    D. 600,000 + 5,000 + 200 + 10 + 1

11. Jaden draws an angle with a measure greater than 0° and less than 90°. What type of angle did he draw?

   A. acute

   B. obtuse

   C. perpendicular

   D. right

12. Draw the lines of symmetry in the figure.

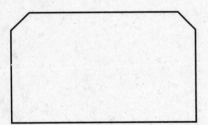

ONLINE TESTING
On the test, you may be asked to draw lines of symmetry on a figure in a grid using a computer. In this book, you will draw the lines by hand.

13. Kai's favorite library is on State Street, which runs between Ohio Avenue and Michigan Road. If State Street intersects Ohio Avenue at an 80° angle and Michigan Road at a right angle, how would you describe the lines that State Street and Michigan Road make?

   A. acute lines

   B. perpendicular lines

   C. obtuse lines

   D. parallel lines

14. Look at the table below.

    Part A: Complete the table.

Input (b)	Output (g)
3	6
4	8
7	14
	34
	82

    Part B: Circle the equation that describes the pattern.

    A. $b + 5 = g$

    B. $g + 5 = b$

    C. $b + 7 = g$

    D. $b \times 2 = g$

15. Circle all of the fractions that make the equation true when substituted for x.

    $$\frac{3}{9} = x$$

    A. $\frac{2}{3}$

    B. $\frac{1}{3}$

    C. $\frac{4}{12}$

    D. $\frac{1}{6}$

16. Kenzie walks 0.65 mile to the soccer fields. Armani walks 0.58 miles to the soccer fields. Who walks farther? Explain.

17. **Part A:** Estimate.

Part B: Solve the problem. Is the actual answer *greater than* or *less than* the estimate? Explain.

18. There are 150 newspapers in a bundle. Each neighborhood or apartment building gets a bundle. There are 22 neighborhoods and apartment buildings. How many newspapers are there?

Part A: Circle all the choices that show a correct way to solve the problem.

A. 150 × 22 =

B. 150 × 20 + 150 × 2 =

C. 150 + 20 × 150 + 2 =

D. 22 × 150 =

Part B: Solve the problem.

_____

> **ONLINE TESTING**
> On the actual test, you may have to click on multiple bubbles. In this book, you will circle the answers.

19. Divide. Show your work. Use multiplication and addition to check.

8)5,802

20. Divide. Use the Distributive Property and the area model to show your work.

372 ÷ 6 =

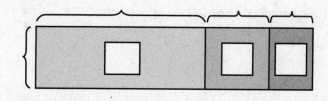

____ + ____ + ____ = ____

372 ÷ 6 = ____

21. Draw the next figure in the pattern.

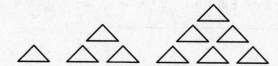

22. Complete the conversion table.

Cups (c.)	Quarts (qt.)	(c., qt.)
	1	
	4	
	6	
	7	

23. Complete the conversion table.

Centimeters (cm)	Kilometers (km)	(cm, km)
	1	
	2	
	5	
	8	

24. Choose the best estimate for the height of a door.

   A. 3 centimeters

   B. 6 kilometers

   C. 3 meters

   D. 6 meters

25. Dante is putting a fence around his garden. The garden will be 2 feet wide and 8 feet long.

   Part A: What is the perimeter of the garden? Explain.

   Part B: What is the area of the garden? Explain.

26. Cole and six of his friends cut some grapefruits into halves.

   Part A: If each person takes $\frac{1}{2}$ of a grapefruit, how many grapefruits would they have taken together? Explain by writing an equation.

   Part B: If the friends cut 5 grapefruits, how many halves were left over after they each took their shares? Explain by writing equations.

   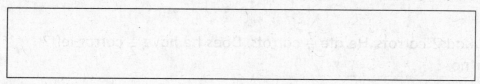

   Part C: Place a point on the number line for the fraction that represents the total amount of grapefruit taken.

   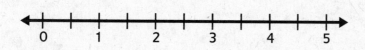

   > **ONLINE TESTING**
   > On the test, you might click on the number line to answer the question. In this book, you will write on the number line.

27. Look at the rectangle below.

   Part A: Shade $\frac{5}{8}$ of the rectangle.

   Part B: Write an equivalent fraction.

Grade 4 • Benchmark Test 4   201

28. Circle all of the fractions that are equal to the sum.

$\frac{5}{6} + \frac{4}{6} =$

A. $\frac{3}{2}$

B. $1\frac{1}{3}$

C. $1\frac{2}{6}$

D. $\frac{9}{6}$

29. Muhammad had 2 carrots. He ate $\frac{7}{4}$ carrots. Does he have $\frac{1}{2}$ carrot left? Circle yes or no.

Yes   No

30. Shade the grid model to show seventy-two out of a hundred.

   Part A: Shade the model.

   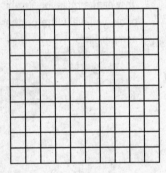

   Part B: Write seventy-two out of a hundred as a decimal.

   Part C: Write seventy-two out of a hundred as a fraction in simplest form.

31. Asia spends three hours biking. How many minutes does she bike? Write an equation.

Performance Task

At the Farm

Maddie and Blaine are helping out at the farm. They are weighing the baby goats.

Write your answers on another piece of paper. Show all your work to receive full credit.

Part A

Maddie and Blaine will use a line plot to record the weights of the goats. Use the measurements of weight in the table to make a line plot.

Baby Goat Weights in pounds				
6	$7\frac{3}{4}$	10	12	$5\frac{1}{2}$
$10\frac{1}{2}$	8	8	$10\frac{1}{2}$	$6\frac{3}{4}$
$9\frac{1}{2}$	$7\frac{3}{4}$	10	11	8

Part B

Look at the line plot. How much does the heaviest goat weigh? How much does the lightest goat weigh? Which weight is plotted the most?

Grade 4 • Benchmark Test 4

Performance Task (continued)

Part C

What is the difference in size between the heaviest goat and the lightest goat? Explain.

Part D

If a goat weighs 8.35 pounds, how would you write that as a fraction? Show the number on a number line.

Part E

If one goat weighs 8.35 pounds and another goat weighs 8.58 pounds, which is heavier? Show your work as a comparison using >, <, or =. Use a number line to show how you found the answer.

Smarter Balanced Assessment Item Types

Assessment Item Types

In the spring, you will probably take a state test for math that is taken on a computer. The problems on the next few pages show you the kinds of questions you might have to answer and what to do to show your answer on the computer.

Selected Response means that you are given answers from which you can choose.

Selected Response Items

Multiple Choice

Regular multiple choice questions are like tests you may have taken before. Read the question and then choose the one best answer.

> **ONLINE EXPERIENCE** Click on the box to select the one correct answer.
>
> **HELPFUL HINT** Only one answer is correct. You may be able to rule out some of the answer choices because they are unreasonable.

A teacher gives 8 students some marbles to play a game. She has 70 marbles total. The teacher gives each student 1 marble until all 70 marbles are gone. How many students get exactly 9 marbles?

- ☐ 2
- ☐ 3
- ☐ 4
- ☒ 6

▶ **Try On Your Own!**

Amy uses $\frac{3}{4}$ of a yard of fabric for each pillowcase that she makes. How many yards of fabric will she need in order to make 9 pillow cases?

- ☐ $5\frac{1}{4}$
- ☒ $6\frac{3}{4}$
- ☐ $7\frac{3}{4}$
- ☐ $8\frac{1}{4}$

Multiple Correct Answers

Sometimes a multiple choice question may have more than one answer that is correct. The question may or may not tell you how many to choose.

> **ONLINE EXPERIENCE** Click on the box to select it.
>
> **HELPFUL HINT** Read each answer choice carefully. There may be more than one right answer.

Select all the numbers that make this inequality true.

$$2\frac{1}{5} > 1\frac{1}{5} + \square + 1$$

- ☐ $\frac{1}{5}$
- ☐ $\frac{3}{5}$
- ☐ $\frac{5}{5}$
- ☐ $\frac{10}{5}$
- ☐ $\frac{4}{5}$

▶ **Try On Your Own!**

Select all equations that are true.

- ☒ $\frac{4}{10} = 0.4$
- ☐ $\frac{24}{100} = 2.4$
- ☐ $\frac{6}{100} = 0.6$
- ☒ $\frac{5}{100} = 0.05$

Grade 4 • Smarter Balanced Assessment Item Types

Smarter Balanced Assessment Item Types

You may have to choose your answer from a group of objects.

Click to Select

🖱 **ONLINE EXPERIENCE** Click on the item to select it.

💡 **HELPFUL HINT** On this page you can draw a circle or a box around the item you want to choose.

Brayden and Ann both collect coins. Brayden has 86 coins. Ann has four times as many coins as Brayden. How many coins does Ann have? Select all the equations that represent this problem.

86 ÷ 4 = ☐   (4 × 86 = ☐)

4 × ☐ = 86   (☐ ÷ 4 = 86)

(☐ ÷ 86 = 4)   86 ÷ ☐ = 4

Try On Your Own!
Select all figures that have line symmetry.

Another type of question asks you to tell whether the sentence given is true or false. It may also ask you whether you agree with the statement, or if it is true. Then you select yes or no to tell whether you agree.

Multiple True/False or Multiple Yes/No

🖱 **ONLINE EXPERIENCE** Click on the box to select it.

💡 **HELPFUL HINT** There is more than one statement. Any or all of them may be correct.

$\frac{4}{6}$ of the rectangle is shaded gray.

Decide if each fraction is equal to $\frac{4}{6}$. Select Yes or No for each fraction.

Yes	No	
☐	☐	$\frac{6}{9}$
☐	☐	$\frac{7}{12}$
☐	☐	$\frac{1}{2}$
☐	☐	$\frac{2}{3}$

Try On Your Own!
Select True or False for each comparison

True	False	
☐	☐	$\frac{5}{6} > \frac{3}{5}$
☐	☐	$\frac{4}{10} > \frac{5}{8}$
☐	☐	$\frac{4}{6} > \frac{3}{4}$
☐	☐	$\frac{2}{3} > \frac{7}{9}$

206   Grade 4 • Smarter Balanced Assessment Item Types

Smarter Balanced Assessment Item Types

Sometimes you must use your mouse to click on an object and drag it to the correct place to create your answer.

Drag and Drop

Drag one fraction to each box to make the statements true.

$2\frac{1}{3}$ = $\boxed{\frac{7}{3}}$

$4\frac{1}{3}$ = $\boxed{\frac{13}{3}}$

$2\frac{1}{3}$ $\boxed{\frac{10}{3}}$ $\boxed{\frac{7}{3}}$ $\boxed{\frac{13}{3}}$

ONLINE EXPERIENCE You will click on a number and drag it to the spot it belongs.

HELPFUL HINT Either draw a line to show where the number goes or write the number in the blank.

Try On Your Own!

Order from least to greatest by dragging each fraction to a box.

$\frac{4}{6}$ $\frac{1}{3}$ $\frac{7}{8}$ $\frac{4}{4}$ $\frac{3}{4}$ $\frac{1}{2}$

$\frac{1}{3}$ $\frac{1}{2}$ $\frac{4}{6}$ $\frac{3}{4}$ $\frac{7}{8}$ $\frac{4}{4}$

When no choices are given from which you can choose, you must create the correct answer. These are called **constructed-response** questions.

One way to answer is to type in the correct answer.

Constructed-Response Items

Fill in the Blank

A pattern is generated using this rule: Start with the number 46 as the first term and subtract 7 from each term to find the next. Enter numbers into the boxes to complete the table.

Term	Number
First	46
Second	39
Third	32
Fourth	25
Fifth	18

ONLINE EXPERIENCE You will click on the space and a keyboard will appear for you to use to write the numbers and symbols you need.

HELPFUL HINT Be sure to provide an answer for each space in the table.

Try On Your Own!

Jorge and his two brothers each water the plants. Jorge's watering can holds half as much water as his older brother's watering can. His younger brother's watering can holds 6 cups. Altogether the watering cans hold 3 gallons. How many cups does Jorge's watering can hold?

14

Grade 4 • Smarter Balanced Assessment Item Types

Multipart Question

Some questions have two or more parts to answer. Each part might be a different type of question.

Each bag of dried apples has 4 servings. Each bag of dried bananas has 5 servings. How many servings of dried fruit are in 7 bags of dried apples and 3 bags of dried bananas?

Part A: Drag the numbers to the boxes and the symbols to the circles to make an equation to show how many servings of dried fruit there are in all.

☐ 4 ◯ 7 ◯ + ☐ 5 ◯ × ☐ 3 = ☐ 43

◯ −   ◯ ÷
◯ +   ◯ ×

☐ 7   ☐ 28
☐ 15 ☐ 4
☐ 3   ☐ 42
☐ 5   ☐ 43

Part B: Select the correct number of servings of dried fruit Anna has.

■ 43 servings
☐ 42 servings
☐ 15 servings
☐ 18 servings

Try On Your Own!

Mr. and Mrs. Lopez are putting tiles on the floor in their kitchen. They can fit 12 rows of 14 tiles in the kitchen. If each tile costs $3, what is the total cost?

Part A: Choose a sentence to find how many tiles they needed in all.

☐ 12 + 14 = 26 tiles
☐ 12 × 2 + 14 × 2 = 52 tiles
■ 12 × 14 = 168 tiles
☐ 12 × 2 + 14 = 38 tiles

Part B: Complete the sentence below.

Mr. and Mrs. Lopez spent ☐ 504 dollars on tile for the kitchen.

Countdown 20

NAME _____ DATE _____ SCORE _____

Countdown: 20 Weeks

> **ONLINE TESTING**
> On the actual test, you might be asked to drag the correct numbers into the place-value chart. In this book, you will be asked to write the numbers in the chart.

1. Use the clues to complete the place-value chart.

Thousands			Ones		
hundreds	tens	ones	hundreds	tens	ones
3	6	8	5	2	4

The 2 has a value of 2 × 10.
The 3 has a value of 3 × 100,000.
The 4 is in the ones place.
The 5 is in the hundreds place.
The 6 is in the ten thousands place.
The 8 has a value of 8 × 1,000.

2. Circle the statement that shows *one hundred seventy-five thousand, two hundred sixty four* in expanded form.

   A. (100,000 + 70,000 + 5,000 + 200 + 60 + 4) ⟵ circled
   B. 175,264
   C. 264,175
   D. 200,000 + 60,000 + 4,000 + 100 + 70 + 5

3. Compare. Use >, <, or =.

   Part A: 689,674 __>__ 689,476
   Part B: 264,864 __=__ 264,864

4. Circle the answer that shows 546,216 rounded to the nearest ten thousands place.

   A. 500,000
   B. 546,200
   C. 556,216
   D. (550,000) ⟵ circled

5. Order the numbers from greatest to least.

   354,678   354,687
   354,687   354,678
   345,876   345,876

Countdown 19

NAME _____ DATE _____

Countdown: 19 Weeks   SCORE _____

1. **Part A:** Find the unknown.

   $57 + (8 + \underline{}) = (57 + 8) + 9$

   Part B: Circle the addition property it shows.

   A. Associative Property (circled)
   B. Commutative Property
   C. Identity Property
   D. Parentheses Property

2. Draw a line connecting the number to its description.

   1,000 less than 85,678 — 85,875
   200 more than 85,675 — 84,678
   10,000 less than 85,675 — 75,675
   10,000 more than 85,675 — 95,675

   > **ONLINE TESTING**
   > On the actual test, you might be asked to drag and drop the numbers. In this book, you will be asked to write the answers.

3. Circle the answers that show ways of estimating the addition problem below.

   46,417
   + 71,654

   A. 72,000
      + 46,000   (circled)

   B. 70,000
      + 40,000

   C. 71,654
      + 46,417

   D. 71,700
      + 46,400   (circled)

4. Subtract. Use addition to check.

   798,656
   − 465,684
   ─────────
   332,972

   332,972
   + 465,684
   ─────────
   798,656

5. Write the number that completes the number sentence.

   $51,268 − \underline{10,000} = 41,268$

NAME _____ DATE _____ SCORE _____

Countdown: 18 Weeks

1. The table shows the production numbers for a company that manufactures cell phones. The table entry is the number of phones manufactured in the given year.

Year	Number of Phones Manufactured
2010	2,345,000
2011	3,155,000
2012	4,050,000
2013	4,891,000

Part A: Between which two years did the company experience the most growth in the number of cell phones they manufactured? What was the increase?

Between 2011 and 2012 the increase was 895,000 phones.

Part B: Estimate to the nearest ten thousand the number of phones manufactured in the four-year period.

14,450,000 phones

2. Mr. Harrison is saving money to pay for a family vacation in August. The total cost of the trip will be $1,195. In March, he saved $185. In April he saved $412. In May he save $204. In June he saved $175. Use the following numbers to estimate how much Mr. Harrison needs to save in July in order to pay for the vacation.

200   200   400   1,200   200

1,200 − (200 + 400 + 200 + 200) = 200

ONLINE TESTING
On the actual test, you might be asked to drag numbers to the correct location. In this book, you will be asked to write the numbers using a pencil.

3. Select each statement that is a member of the fact family for the array.

☐ 5 × 6 = 30
☐ 5 + 5 = 10
☐ 30 ÷ 6 = 5
☐ 30 ÷ 5 = 6
☐ 30 ÷ 10 = 3

4. Jamal earned $12 the first snowfall of the year for shoveling driveways. During the heaviest snowfall of the year, he earned 4 times the amount that he earned during the first snowfall. Use numbers and symbols from the list shown to write an equation that can be used to find a, the amount that Jamal earned during the heaviest snowfall of the year.

3   4   8   12 a ×   ÷   +   −

$a = 12 \times 4$

5. There are 12 students on a soccer team. Each student has 4 boxes of ornaments to sell for a fundraiser. Each box has 3 ornaments in it. Select all number statements that could be used to find out how many ornaments there are in all.

☐ (3 × 4) × 12
☐ 3 × (4 × 12)
☐ 4 × 12
☐ 3 × 12
☐ 12 × 12
☐ (3 + 4) × 12

Countdown 17

NAME _____ DATE _____

Countdown: 17 Weeks   SCORE _____

> **ONLINE TESTING**
> On the actual test, you may be asked to arrange the objects by clicking on them. In this book, you will instead draw the objects using a pencil.

1. A tray of muffins is arranged as shown. Draw one other way that the same muffins can be arranged into rows and columns

   Sample answer:

2. A grocer has 32 boxes of rice cereal, 18 boxes of oat cereal, and 13 boxes of bran cereal. He wants to put these boxes into three different displays, and he wants each display to have the same number of boxes.

   Part A: Is it reasonable to state that there will be more than 15 boxes in each display? Explain.

   Yes. If there were 15 boxes in each of 3 displays, there would be a total of 45 boxes. Since there are more than 45 boxes, the total in each display is more than 15.

   Part B: Write and solve an equation whose solution is b, the number of boxes in each display.

   $63 \div 3 = b$

3. Two swimmers keep track of the number of laps they swim each day. For each day, tell whether Swimmer A swims 3 times as many laps as Swimmer B.

Day	Swimmer A	Swimmer B	Does A swim 3 times as many laps as B?	
			Yes	No
Monday	12	4	✓	
Tuesday	3	9		✓
Wednesday	18	7		✓
Thursday	15	5	✓	
Friday	9	6		✓

4. Quentin was solving a multiplication problem. Circle the names of all properties he used to solve the problem.

   $5 \times (6 \times 2) = 5 \times (2 \times 6)$

   $ = (5 \times 2) \times 6$

   $ = 10 \times 6$

   $ = 60$

   (Associative Property of Multiplication)   (Commutative Property of Multiplication)   Identity Property of Multiplication

5. The number 42,___2 rounded to the hundreds place is 42,300. What is the least possible sum of the two missing digits? Justify your answer.

   3; Sample answer: The smallest that the hundreds place can be is 2, but then the tens place would have to be at least 5, for a sum of 7. If the hundred place is 3, the tens place can be 0, so the sum would be $3 + 0 = 3$.

Countdown: 16 Weeks

NAME _____ DATE _____

SCORE _____

> **ONLINE TESTING**
> On the actual test, you might be asked to click on the boxes to select them. In this book, you will instead shade the boxes with a pencil.

1. Select each expression that represents the model shown.

   ☐ (5 × 10) + (5 × 6)

   ☐ 5 × 16

   ☐ (5 × 10) + (6 × 10)

   ☐ (5 × 15) + 5

2. The Suarez family is taking a vacation. They will drive 584 miles every day for three days.

   A. Estimate the number of miles the family will drive in total over the three days. Show your estimate.

   $600 \times 3 = 1,800$ miles

   B. Will the exact number of miles be less than or greater than your estimate? Explain.

   It will be less because I rounded 584 up to 600.

   C. Find the exact number of miles the family will drive over the three days.

   1,752 miles

3. Select Yes or No to indicate whether or not the product will require regrouping.

   Requires Regrouping?

Yes	No	
☒	☐	305 × 9
☐	☒	1,234 × 2
☒	☐	1,402 × 3

4. Henry is looking at a set of base ten blocks.

   A. What multiplication problem does this represent?

   37×3

   B. What is the answer to the multiplication problem?

   _____

5. Use the Associative Property of Multiplication to solve the equation in the easiest way possible. Show your work.

   $5 \times (4 \times 14) =$ _____

   $5 \times (4 \times 14) = (5 \times 4) \times 14 = 20 \times 14 = 280$

Countdown 15

NAME _____ DATE _____

SCORE _____

Countdown: 15 Weeks

1. Mrs. Giovanni plants a small garden of tomatoes every year. She places the plants into 26 rows of 15 plants each.

   Part A: Which expression shows the total number of plants?

   A. (26 × 10) + (26 × 5) ⭕
   B. (26 × 10) + (26 × 6)
   C. (15 × 20) + (15 × 5)
   D. (26 + 10) × (26 + 5)

   Part B: Find another arrangement that would result in the same number of plants.

   Sample answer: 10 × 39

2. Estimate each product by rounding the larger number to the greatest place value possible. Then place an X in the appropriate box to indicate whether the estimate is greater than or less than the actual product.

Product	Estimate	Product is ___ than the estimate	
		Less	Greater
205 × 8	1,600		X
1,827 × 5	10,000	X	
3,127 × 4	12,000		X
18,412 × 7	140,000	X	

   ONLINE TESTING On the actual test, you may be asked to click in the boxes to add the X's. In this book, you will instead write the X's using a pencil.

3. Manuel is making cookies for a bake sale. He makes 16 batches. Each batch has *b* cookies in it. The number *b* is an odd 2-digit number. Indicate whether each statement is true or false.

True	False	
☐	☐	The total number of cookies is odd.
☐	☐	The total number of cookies could be less than 100.
☐	☐	The total number of cookies could have a zero in the one's place.
☐	☐	The total number of cookies could be a multiple of 10.
☐	☐	The number equal to 16 × *b* could be greater than 1,000.

4. Which number correctly completes the equation?

   26 × 72 = _____ + 40 + 420 + 12

   A. 140
   B. 1,400 ⭕
   C. 14,000
   D. 140,000

5. Uma scored 8 goals in her first season of soccer and 24 goals in her second season. Write two different statements that relate the two numbers. The first statement should involve addition. The second statement should involve multiplication.

   Addition Statement

   Sample answer: Uma scored 16 more goals in her second season than she did in her first season.

   Multiplication Statement

   Sample answer: Uma scored three times the number of goals in her second season that she did in her first season.

NAME _____ DATE _____

SCORE _____

Countdown: 14 Weeks

ONLINE TESTING
On the actual test, you may be asked to click into the boxes in order to key in the answers. In this book, you will instead write the answers using a pencil.

1. A farmer has a garden that he is dividing into four different regions. The garden can be modeled using an area model as shown. All measurements are in feet.

Part A: Fill in the area of each of the four regions.

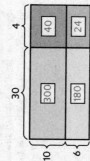

Part B: What is the area of the entire garden?

544 square feet

2. Which of the following number sentences are not useful for finding 24 × 62?

A. 20 × 2 = 40
B. 24 × 2 = 48
C. 24 × 6 = 144
D. 4 × 2 = 8

3. Mrs. Shen is trying to write a multiplication problem for her fourth grade math class. She wants her students to expand two 2-digit numbers being multiplied together. She wants the answer to be:

(20 × 50) + (20 × 6) + (2 × 50) + (2 × 6)

Part A: What should the original multiplication problem be?

22 × 56

Part B: What is the final answer to the problem?

1,232

4. List three different ways of arranging 72 seats for a concert in rows and columns.

Sample answer: 12 × 6, 4 × 18, 3 × 24

5. A marathon runner runs 17 miles every day for training. He runs for d days and runs a total of 119 miles. Circle **all** of the following that express the relationship between 17, d, and 119.

17 × d = 119

17 × 119 = d

d × 119 = 17

⟨119 ÷ d = 17⟩

⟨119 ÷ 17 = d⟩

17 ÷ d = 119

NAME _____ DATE _____

Countdown: 13 Weeks   SCORE _____

ONLINE TESTING
On the actual test, you may be asked to click on the boxes to select each answer. In this book, you will instead shade the boxes using a pencil.

1. An electronics company plans on producing 7,158 components in the next 12 days. The plan manager estimates that they should produce about 600 components per day. Select **each** true statement about the plant manager's calculation.

   ☐ The manager rounded to the nearest hundred before dividing.

   ☐ The manager used compatible numbers to estimate.

   ☐ The manager's estimation can be checked with the product 600 × 12.

   ☐ The manager's estimation can be checked by adding 28 to 7,100.

   ☐ The manager's estimate is less than the actual number of components produced on a given day if they produce the same number of components each day.

2. A baker baked 6 types of bread. He baked the same number of each type. There are 360 loaves of bread total. How many loaves of each type did the baker bake?

   [60 loaves]

3. A house painter has a goal of painting 500 rooms in 8 months. He uses the long division 8)500 to determine how many rooms he needs to paint each month. Determine whether each statement below is true or false.

   True False
   ☐ ☐ His problem has a remainder.
   ☐ ☐ If he paints 63 rooms each month, he will meet his goal.
   ☐ ☐ If he paints 62 rooms each month, he will meet his goal.

4. Four friends raised a total of $732 for a local charity. Each friend raised the same amount of money.

   Part A: Model the total amount that the friends raised using the least number of base-ten blocks possible.

   Part B: Show how to use the base-ten blocks to determine how much each friend raised. You may trade for different base-ten blocks, if necessary.

5. An electrician orders 48 packages of circuit breakers. Each packages contains 17 breakers.

   Part A: Write the correct numbers and symbols in the boxes to make an equation that shows an estimate of the number of circuit breakers ordered. Select from the list below.

   20 30 40 50 200 300 400 500 800 1,000

   + − ÷ ×

   [50] × [20] = [1,000]

   Part B: Is the estimated number of circuit breakers ordered greater than or less than the actual number? Explain how you know.

   Sample answer: The estimate is greater. I rounded 48 up to 50 and 17 up to 20, so the estimate is greater than the actual number.

Countdown 12

NAME _____ DATE _____

SCORE _____

Countdown: 12 Weeks

ONLINE TESTING
On the actual test, you may be asked to drag the steps into order. In this book, you will instead number the steps using a pencil.

1. A woodworker is building 5 tables. He wants to figure out how many feet of wood each table can have. He has 603 feet of board. The woodworker will check his work. Write the numbers 1-3 next to the correct three steps below to show the order in which the woodworker should perform them.

   ___ Add 2 to 601.
   ___ Add 3 to 120.
   ___ Divide 120 by 5.
   _1_ Divide 603 by 5.
   ___ Multiply 121 by 5.
   _2_ Multiply 120 by 5.
   _3_ Add 3 to 600.
   ___ Subtract 3 from 600.

2. A florist orders 820 roses for a reception. Four of the roses were too damaged to use. The rest of the roses will go into vases holding 8 roses each.

   Part A: Write an equation that can be used to find v, the number of vases that the florist will need.

   $816 \div 8 = v$

   Part B: Think about the numbers in the division problem. Explain how you can solve the problem using place value and mental math. Then state the answer.

   Sample answer: I have 8 hundreds, and 16 extra. If I divide each by 8, I get 1 hundred, and 2 extra. So, there will be 102 vases.

3. A grocer is ordering juice containers that come in packages that have 7 bottles. The grocer starts writing out multiples of 7 in order to place a juice order. Select all correct statements.

   ☐ All of the numbers on the list are odd.
   ☑ The first ten multiples will contain every possible digit in the one's place before they start repeating.
   ☐ The sum of the digits in each number in the list is divisible by 7.
   ☑ Each number in the list is divisible by 7.
   ☐ Each number in the list is divisible by 14.

4. Xavier practices 10 free throws on Monday, 20 on Tuesday, 40 on Wednesday, and 80 on Thursday.

   Part A: Describe the pattern.

   Sample answer: The next day can be determined by multiplying the previous day by 2.

   Part B: Describe why this pattern is not reasonable for Xavier to continue by looking at the next three days.

   Sample answer: If the pattern continues, then Xavier will shoot 160 on Friday, 320 on Saturday, and 640 on Sunday. The number increases very rapidly, and Xavier will not have the time to shoot all of the free throws.

5. The first term in a number pattern is 20. Additional terms are obtained by multiplying the previous term by 4. What is the first term that is larger than 1,000? Enter numbers into the table to figure this out. Stop when you get a term greater than 1,000.

Term	Number
First	20
Second	80
Third	320
Fourth	1,280
Fifth	

Countdown 11

NAME _____ DATE _____ SCORE _____

Countdown: 11 Weeks

1. In the table shown, each rule's input is the previous rule's output.

Part A: Complete the table by writing in the missing values.

Input	Add 3	Subtract 5	Multiply by 5
5	8	3	15
6	9	4	20
15	18	13	65
8	11	6	30

Part B: If n is the input, write an equation that can be used to find p, the output.

$(n + 3 - 5) \times 5 = p$

2. A scientist is studying an ant infestation at a local building site. On the first day, he observes 16 ants along the base of the entryway. On the second day, he observes 20 ants. On the third, he observes 24 ants. This pattern continues for the rest of the week. Indicate whether each statement is true or false.

ONLINE TESTING On the actual test, you may be asked to click into the boxes to shade them. In this book, you will instead shade the boxes using a pencil.

True	False	
☐	☐	The number of ants is always a multiple of 4.
☐	☐	The rule for the number of ants seen the next day is "add 8".
☐	☐	The number of ants is always a multiple of 8.
☐	☐	The number of ants is always a multiple of 2.
☐	☐	The total number of ants for the week is a multiple of 4.

3. How many branches will be at the bottom of the 7th figure in the pattern shown?

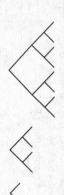

128

4. A large pot of water contains 64 cups of water. As the water boils, the amount of water in the pot decreases. After 10 minutes, there are 48 cups left. After another 10 minutes there are only 32 cups left. If this pattern continues, circle the number of minutes it will take until all of the water is gone.

A. 4 minutes   C. 30 minutes

B. 40 minutes   D. 50 minutes

5. A store is putting together packages of school supplies to donate to a local school. The table shows the number of each type of each supply they have to donate.

Type of Supply	Number of Supply
Pencils	172
Pens	161
Packs of Paper	45
Book Bags	31

Part A: How many boxes can be packed if each package must contain 10 writing utensils (pens or pencils)?

33 packages

Part B: If each package will be put into a book bag, are there enough book bags to fit your answer from *Part A*? Explain.

No. Sample answer: There are only 31 book bags, so the store cannot make 33 packages.

Countdown 10

NAME _____ DATE _____ SCORE _____

Countdown: 10 Weeks

ONLINE TESTING
Sometimes you can take values directly from a table as the solution to your problem. In other cases, you may need to take the values from the table and modify them before you can find your solution.

1. Dr. Nicole Soto has developed Ivy-X, a poison ivy remedy that she thinks works better than the standard treatment. Dr. Soto's data is shown.

	Got better	Got worse
Ivy-X treatment	12	4
Standard treatment	10	5

Part A: What fraction of the patients who used Ivy-X got better? What fraction got worse?

Better: $\frac{12}{16} = \frac{3}{4}$, worse: $\frac{4}{16} = \frac{1}{4}$

Part B: What fraction of the patients who used the standard treatment got better? What fraction got worse?

Better: $\frac{10}{15} = \frac{2}{3}$, worse: $\frac{5}{15} = \frac{1}{3}$

Part C: Is Dr. Soto correct? Does Ivy-X actually work better than the standard poison ivy treatment? Explain.

Sample answer: $\frac{3}{4}$ is greater than $\frac{2}{3}$ so Ivy-X seems to work slightly better than the standard treatment.

2. Ricky planned to spend $20 to enter Sky Rock Water Park and buy extra tickets for 4 special rides while he is there. The special rides cost $3 each. Ricky wrote this expression to show the total amount of money he needs to bring.

$$20 + 4 \times \$3$$

Using the expression above, Ricky calculates that he needs to bring a total of $72 to the park. Is Ricky correct? Explain. If he is not correct, how much does he need to bring?

No; Ricky did not use order of operations for his calculation; $32

3. Craig told Dava the first two numbers in his locker combination. To find the third number Craig said, "Find the sum of all of the factors of the number 20."

What are the factors of the number 20? What is the third number of Craig's combination?

1, 20, 2, 10, 4, 5; 42

4. Explain why the model does or does not show $\frac{3}{8}$.

The model does not show $\frac{3}{8}$ because, though 3 of 8 sections are shaded, the sections are not equal in size.

5. What mixed number is equivalent to $\frac{24}{5}$? How can you use models to show that your answer is correct? Explain.

$4\frac{4}{5}$; Sample answer: My model would show 4 wholes, each divided into 5 equal parts with all of the parts shaded in. Then I would have another whole divided into 5 equal parts with 4 of the 5 parts shaded. Counting the fifths would show that I have 4×5 fifths, plus 4 more fifths, a total of $\frac{24}{5}$.

Countdown 9

NAME _____ DATE _____

SCORE _____

Countdown: 9 Weeks

1. The attendance at a town hall meeting is expected to be 75. Folding chairs for the meeting are arranged in rows of 6.

Part A: How many rows are needed to fit all of the people?

13

Part B: If exactly 75 people show up for the meeting, how many seats will be left empty?

3

2. Between 0 and 99, the greatest number of prime numbers end in which digit — 0, 1, 2, 3, 4, 5, 6, 7, 8, or 9? Explain.

Prime numbers from 0–99 are: 2, 3, 5, 7, 11, 13, 17, 19, 23, 29, 31, 37, 41, 43, 47, 53, 59, 61, 67, 71, 73, 79, 83, 89, 97. There are 7 prime numbers ending in 3, which is the most of any digit.

ONLINE TESTING
For problems like this one you might find it helpful to use the problem-solving strategy of make a list. Use this strategy whenever you have a group of items from which you need to find a pattern.

3. Kalief claims that he can carry out any mathematical operation forward or backward. As an example, Kalief shows the equations below. Is Kalief correct? Explain.

7 + 4 = 11
4 + 7 = 11

Kalief is not correct. Though the Commutative Property works for both addition and multiplication, it does not work for subtraction or division. For example, 7 − 4 does not equal 4 − 7. Similarly, 12 ÷ 3 does not give the same quotient as 3 ÷ 12.

4. In basketball, a player is considered an "excellent" shooter if he or she makes more than half of his or her shots.

Player	Shots made	Shots taken
Steve	10	16
Boris	12	24
Kim	15	35
Shaniqua	25	40

Part A: Which of these players would be considered an excellent shooter? Explain.

Steve and Shaniqua; Boris made half of his shots, but to be considered excellent, a player needs to make more than half of his or her shots.

Part B: Who was the best shooter in the group? Explain.

Both Steve and Shaniqua made the same fraction of their shots, $\frac{5}{8}$, so they tied for being the best shooter.

5. Explain why each model does or does not show $\frac{8}{3}$

A.

B.

C.

D.

A. The model shows $\frac{8}{3}$ because it has 2 wholes plus $\frac{2}{3}$ of another whole.

B. The model does not show $\frac{8}{3}$ because it has 2 wholes plus $\frac{3}{4}$, not $\frac{2}{3}$ of another whole.

C. The model does not show $\frac{8}{3}$ because it shows $\frac{8}{9}$, not $2\frac{2}{3}$.

D. The model shows $\frac{8}{3}$ because if you move the unshaded parts together, you have 2 wholes and $\frac{2}{3}$ of another whole.

Countdown 8

NAME _____ **DATE** _____

SCORE _____

Countdown: 8 Weeks

> **ONLINE TESTING**
> On the actual test, you may need to click on the screen to show a location on the diagram. In this book, you will use your pencil to circle the location.

1. Every day Rafael runs around the 1-mile track that is divided into 8 sections of equal length.

   Part A: From the starting line Rafael ran $\frac{11}{8}$ of a mile in the direction shown. Circle the line that shows the location on the track where Rafael ended his run.

   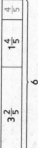

   Start

   Part B: From where Rafael stopped, how much farther does he need to run to reach the starting line again?

   $\frac{5}{8}$ of a mile

2. Bonnie put 5 gold coins that each weighed $\frac{2}{7}$ of a pound into a cloth sack. To the sack Bonnie added 4 silver coins that weighed $\frac{3}{7}$ of a pound each. What was the total weight of the coins in the sack? Explain how you found this total.

   $3\frac{1}{7}$ pounds; Sample answer: I multiplied $\frac{2}{7}$ by 5 to get $\frac{10}{7}$ for the gold coins. I multiplied $\frac{3}{7}$ by 4 to get $\frac{12}{7}$ for the silver coins. Then I added $\frac{10}{7} + \frac{12}{7}$ to get $\frac{22}{7}$ for the total. Simplifying gave $3\frac{1}{7}$ pounds.

3. In the election for mayor, Victor Peña got 142,512 votes and Marley Robins received 142,483 votes.

   Part A: Who won the election and by how much?

   Peña won by 29 votes.

   Part B: The election rules state that all results must be rounded to the nearest hundred. According to the rule, who won the election? Explain.

   According to the rule, the election is a tie. Both candidates got 142,500 votes when rounded to the nearest hundred.

4. Petra's grandfather lives 6 miles from Petra's home. From home, Petra rode her bike $3\frac{2}{5}$ miles toward her grandfather's house and stopped at her friend Hasan's. From Hasan's, Petra rode another $1\frac{4}{5}$ miles and stopped to see her friend Alfie. At Alfie's, how far from is Petra from her grandfather's house? Use the diagram to help explain how you got your answer.

$3\frac{2}{5}$	$1\frac{4}{5}$	$\frac{4}{5}$

   6

   $\frac{4}{5}$ mile; Sample answer I added $3\frac{2}{5}$ and $1\frac{4}{5}$ to get $5\frac{1}{5}$. Then I subtracted $5\frac{1}{5}$ from 6, the total distance, to get $\frac{4}{5}$ mile left.

5. Which of following are equivalent to $4\frac{3}{8}$? Select all that apply.

   ☒ $7 \times \frac{5}{8}$
   ☐ $6 - 2\frac{5}{8}$
   ☐ $3\frac{5}{8} + \frac{6}{8}$
   ☒ $35 \times \frac{1}{8}$

Countdown 7

NAME _____ DATE _____ SCORE _____

Countdown: 7 Weeks

1. A radio station runs three different types of commercials each hour. The commercials each last a different amount of time.

Commercial type (minutes)	Short ($\frac{1}{6}$ min)	Medium ($\frac{1}{4}$ min)	Long ($\frac{2}{3}$ min)
Commercials per hour	10	12	8
Total time (min)	$1\frac{2}{3}$	3	$5\frac{1}{3}$

Part A: How many minutes does each commercial type take up per hour on the station? Complete the table.

Part B: The station manager claims that the station runs less than 10 minutes of commercials each hour. Is she correct? Explain.

The manager is not correct. $1\frac{2}{3} + 3 + 5\frac{1}{3} = $ 10 minutes, which is not less than 10.

2. In the right column, write the values from the box that are equivalent to each expression in the left column of the table.

	Box
$4 \times \frac{4}{9}$	$\frac{16}{9}$, $8 \times \frac{2}{9}$
$\frac{12}{5}$	$4 \times \frac{3}{5}$, $12 \times \frac{1}{5}$
$5 \times \frac{2}{3}$	$10 \times \frac{1}{3}$, $3\frac{1}{3}$

Box: $10 \times \frac{1}{3}$, $4 \times \frac{3}{5}$; $12 \times \frac{1}{5}$, $\frac{16}{9}$, 9; $3\frac{1}{3}$, $8 \times \frac{2}{9}$

ONLINE TESTING On the actual test, you may drag items to solve a problem. In this book, you will use your pencil to write in the items.

---

3. Allie picked 8 pounds of strawberries, which was far more than she or her family could eat. Allie gave $1\frac{3}{10}$ pounds to Dov, who said he was going to make a pie. Allie gave $2\frac{9}{10}$ pounds to Nina, who was making strawberry shortcake. After giving away strawberries to Dov and Nina, how many pounds of strawberries did Allie have left? Explain.

$3\frac{4}{5}$ pounds; Sample answer: I added $1\frac{3}{10} + 2\frac{9}{10}$ to get $4\frac{1}{5}$. Then I subtracted $4\frac{1}{5}$ from 8 to get $3\frac{4}{5}$.

4. Tyler collected 8 butterflies. Flor collected 3 times as many butterflies as Tyler. Misha collected 5 times as many butterflies as Flor. How many more butterflies did Misha collect than Tyler?

Misha collected 120 butterflies, which is 112 more than Tyler.

5. What product does this model represent? Fill in the labels for the model and explain how the model helps you find the product.

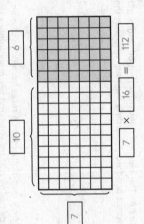

Labels: 10, 6, 7, $7 \times 16 = 112$

The diagram shows $7 \times 16 = 112$. It uses the distributive property: $7 \times 16 = 7 \times (10 + 6) = (7 \times 10) + (7 \times 6) = 70 + 42 = 112$.

Countdown: 6 Weeks

NAME _____ DATE _____

SCORE _____

ONLINE TESTING
On the actual test, you may drag items to show your answer. In this book, you will use your pencil to write in the items.

1. The five judges at the skateboard competition each gave out the scores to Tracee as shown. Write each score into the table from low to high.

   Judge 1: $\frac{9}{10}$ Judge 4: $\frac{27}{100}$
   Judge 2: $\frac{9}{100}$ Judge 5: 0.08
   Judge 3: 0.42

   To get a final score, the high and the low scores are thrown out. The middle three scores are then added to give the final score. What was Tracee's final score? Write your answer as a decimal.

5 (low)	4	3	2	1 (high)
0.08	$\frac{9}{100}$	$\frac{27}{100}$	0.42	$\frac{9}{10}$

   Tracee's final score: 0.78

2. Which models correctly show the sum of $\frac{30}{100} + 0.2$? Select all that apply.

3. Results from the Catch and Release Bass Fishing Contest are shown.

Name	Doug	Kendal	Monk	Daphne
Length (m)	0.54	$\frac{6}{10}$	$\frac{9}{100}$	0.5

   Who was the winner of the contest? Explain how you know.

   Kendal was the winner because $\frac{6}{10}$ is equal to 0.6, which is greater than 0.54, $\frac{9}{100}$, or 0.5.

4. Renting a canoe costs $6 per hour. All renters must also pay $3 for the paddle and life jacket. Zern wrote this equation for finding t, the cost of renting a canoe, for h hours.

   $(6 \times h) + 3 = t$

   Find the rental cost, t, when $h = 4$ hours. Show your work.

   The total cost, t, is $27.
   $(6 \times h) + 3 = t$
   $(6 \times 4) + 3 = t$
   $24 + 3 = t$
   $27 = t$

5. Bruce has measured his feet to be between 0.2 and 0.21 meters in length. Carrie has a pair of red bowling shoes that measure $\frac{23}{100}$ of a meter and and blue pair that are $\frac{18}{100}$ m in length. Can Bruce fit into either pair of Carrie's bowling shoes? Explain.

   Sample answer: Bruce can fit into the red pair because $\frac{23}{100}$ is greater than 0.21. The blue pair is too small; $\frac{18}{100}$ m < 0.2 m.

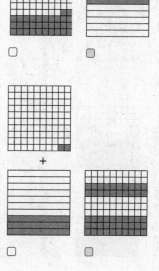

Countdown 5

NAME _____ DATE _____

SCORE _____

Countdown: 5 Weeks

1. Look at the following model.

0.3 + 0.64 = 0.94

Part A: Write a problem that would require you to use this model to find its solution.

Sample answer: Jory poured $\frac{3}{10}$ of a gallon of water into a bucket. Cory poured 0.64 of a gallon into the same bucket. How many gallons of water were in the bucket?

Part B: Solve the problem you wrote. Include a diagram and labels.

0.3 + 0.64 = 0.94

ONLINE TESTING
On the actual test, you may drag items to show your answer. In this book, you will use your pencil to write in the items.

2. Ursula sold 47 copies her new app for $26 each. How much money did Ursula make? Use the numbers below to fill in the area model. You may not use all of the numbers.

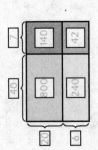

800   36   42   6   20
400   140   7   24   240

3. Ruby has read 0.85 of a book. What fraction of the book does Ruby have left to read? Write your answer in simplest form.

$\frac{3}{20}$

4. It snowed five times in December with snowfalls of 7 inches, 2 inches, 6 inches, 11 inches, and 5 inches. Use the information below to fill in the table.

- The smallest snow amount was on December 16.
- The greatest snow amount was on December 30.
- The snow on December 28 was less than the snow on December 14.
- The snow on December 5 was less than the snow on December 28.

Date	Dec 5	Dec 14	Dec 16	Dec 28	Dec 30
Snow (inches)	5	7	2	6	11

5. Which of the following is equivalent to $\frac{4}{10}$? Select all that apply.

☐ 0.25 + 0.15
 $\frac{2}{5}$
☐ 0.1 + 0.03
☐ $\frac{12}{100}$ + 0.28

Countdown 4

NAME _____ DATE _____ SCORE _____

Countdown: 4 Weeks

ONLINE TESTING
On the actual test, you may drag items to show your answer. In this book, you will use your pencil to write in the items.

1. Oscar figured out a way to multiply 34 × 18 using the Distributive Property and subtraction rather than addition. Oscar's idea is to multiply 34 by the "easy" number of 20 rather than 18. Then, to make up for adding 2 to 18 to make 20, Oscar will subtract 34 times 2 from the total. Fill in the blanks to show Oscar's method.

34 × 18 = (__34__ × 20) − (34 × __2__)
= (__680__) − (__68__)
= __612__

2. Sundip conducted a survey to see how much time each classmate spends per week playing video games. Use Sundip's data to create a line plot. Make sure you include a scale for your line plot.

Less than 1 hr	1 to 2 hrs	2 to 3 hrs	3 to 4 hrs	4 to 5 hrs	More than 5 hrs
4	2	6	7	3	2

Hours

Grade 4 · Countdown 4 Weeks   **45**

3. Ned is building a wooden table. The table top has three sections and a total width of 8 feet. Ned has made the two outer pieces that measure 30 inches and 28 inches in width. How wide should the middle section be? Explain.

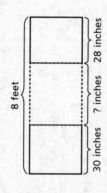

38 inches; The entire width is 8 × 12 = 96 inches. The 2 outer pieces have a sum of 30 + 28 = 58 inches. So, total − known pieces = unknown piece, or 96 − 58 = 38 inches.

4. Which of the following is equivalent to 9 pints? Select all that apply.
  - ☐ 172 fluid ounces
  - ☐ 1 gallon and 12 fluid ounces
  - ☐ 3 quarts and 6 cups
  - ☑ 1 gallon and 1 pint

5. Without shoes, Morgan steps on a scale and weighs exactly 65 pounds. Morgan puts on a pair of sneakers and holds a basketball that weighs 23 ounces. Now she weighs $68\frac{1}{4}$ pounds. How much does each shoe weigh? Show your work.

Use guess and check:
$68\frac{1}{4} - 65 = 3\frac{1}{4}$ pounds = 52 ounces of extra weight
$x + x + 23 = 52$
$12 + 12 + 23 = 47$ [too low]
$15 + 15 + 23 = 53$ [close, but too high]
$14 + 14 + 23 = 51$ [close, but too low]
$14\frac{1}{2} + 14\frac{1}{2} + 23 = 52$ [correct]

So each shoe weighs $14\frac{1}{2}$ ounces.

46   *Grade 4 · Countdown 4 Weeks*

Countdown 3

NAME _____ DATE _____

Countdown: 3 Weeks   SCORE _____

> **ONLINE TESTING**
> On the actual test, you will click on answers to multiple choice questions. In this book, you may be asked to circle answers to multiple choice questions.

1. The Z-Dog company is conducting a survey to help design its new tablet, the Z-Six. The Z-Six is designed for commuters who travel long distances to school or work each day. Circle the correct units for each question in Z-Dog's survey.

   A. What approximate length should the Z-Six tablet have?

   millimeters   (centimeters)   meters   kilometers

   B. What exact width should the Z-Six tablet have?

   (millimeters)   centimeters   meters   kilometers

   C. How far do you travel to school or work?

   millimeters   centimeters   meters   (kilometers)

   D. What is the distance from your desk to the nearest electrical outlet?

   millimeters   centimeters   (meters)   kilometers

2. Jillian deposited a total of $120 in the bank over three days.

   Part A: On Tuesday Jillian deposited $\frac{2}{5}$ of the total. How much did she deposit on Tuesday?

   $\frac{2}{5} \times 120 = \$48$

   Part B: On Wednesday, Jillian deposited half of the total. How much did Jillian deposit on Wednesday?

   $\frac{1}{2} \times 120 = \$60$

   Part C: On Thursday, Jillian deposited the rest of the money. How much did Jillian deposit on Thursday? Explain.

   Jillian deposited $12 on Thursday.
   Total − Tuesday − Wednesday = Thursday
   $120 − $48 − $60 = $12

3. Nina wants to hang this picture so that the amount of space on each side of the picture is the same. How far from the end of the wall should Nina hang the picture? Show your work.

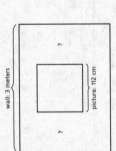

wall: 3 meters
picture: 112 cm

Sample answer: I used guess and check. The distance is 94 cm.
3 m = 300 cm
300 cm = x + x + 112
= 90 + 90 + 112 = 292 (too small)
= 95 + 95 + 112 = 302 (too big)
= 94 + 94 + 112 = 300 (correct)

4. Which of the following is equivalent to 2.04 liters? Select all that apply.

   ☐ 204 milliliters
   ☐ 2 liters and 4 milliliters
   ☒ 2 liters and 40 milliliters
   ☒ 2,040 milliliters

5. The quotient of 350 ÷ 10 is equal to what number divided by 1,000?

   35,000

Countdown 2

NAME _____ DATE _____
SCORE _____

Countdown: 2 Weeks

1. Gina is building a garage that needs to have 96 square meters of space. The garage must have a width of at least 4 meters and be measured in whole meters.

   Part A: How many different plans can Gina use for the garage? List the dimensions of each plan.

   3 plans; 4 m × 24 m, 6 m × 16 m, 8 m × 12 m

   Part B: To have the smallest possible perimeter, which plan should Gina choose? Explain.

   Sample answer: 8 m × 12 m has a perimeter of 40 m, less than 56 m for the 4 m × 24 m plan, and 44 m for the 6 m × 16 m plan.

2. A rectangle has a length of 15 feet and a perimeter of 42 feet. Which of the following is true? Select all that apply.

   ☐ The figure is a square.
   ☐ The width of the rectangle is 12 feet.
   ▪ The area of the rectangle is 90 square feet.
   ▪ One side of the rectangle measures 6 feet.

   ONLINE TESTING
   On the actual test, you will click on answers to multiple choice questions. In this book, you will color in boxes to answer multiple choice questions.

3. Diego explained how he solved the problem below. First he subtracted 30 from 54 to get 24. Then he divided 24 by 3 to get 8 and then added 4 to 8 to get a final answer of 12. Is Diego's solution correct? Explain.

   $$54 - 30 \div 3 + 4$$

   No, Diego should have divided 30 ÷ 3 first to get 10, then subtracted 10 from 54 to get 44. Finally, he should have added 4 to 44 to get 48.

4. If you continue the pattern of shaded squares surrounded by unshaded squares, at what point in the pattern does the area of the outer unshaded squares become greater than the area of the unshaded squares? Explain.

   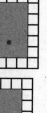

   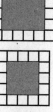
   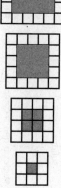

   In the 7 × 7 square the shaded squares have an area of 25 square units, while the area of the outer unshaded squares is 24 square units.

5. Milton started the morning with a full gallon of milk. He used 3 cups of milk to make biscuits and 3 pints of milk to make pudding. To make pancakes, Milton needs 6 cups of milk. Does he have enough milk to make the pancakes? Explain.

   Yes, 1 gallon = 16 cups. Milton used 3 cups + 3 pints = 9 cups. If he uses 6 cups to make pancakes, he will have 1 cup left.

Countdown 1

NAME _____ DATE _____

SCORE _____

Countdown: 1 Week

ONLINE TESTING: On the actual test, you will use your cursor to draw lines. In this book, you will draw lines using your pencil.

1. The octagon below has 8 sides that are equal in length.

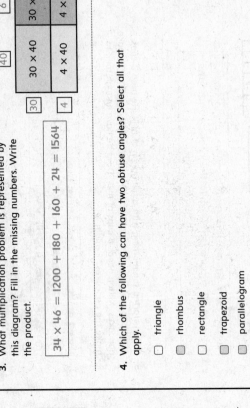

Part A: How many lines of symmetry does the figure have? Draw them.

8 lines of symmetry

Part B: How many equal-sized triangles do the lines of symmetry form?

16 triangles

2. Match the numbers on the left to their factors on the right.

40 — 8,5
72 — 12,6
56 — 4,14
42 — 3,14

3. What multiplication problem is represented by this diagram? Fill in the missing numbers. Write the product.

	30	4
40	30 × 40	4 × 40
6	30 × 6	4 × 6

$34 \times 46 = 1200 + 180 + 160 + 24 = 1564$

4. Which of the following can have two obtuse angles? Select all that apply.

☐ triangle
■ rhombus
☐ rectangle
■ trapezoid
■ parallelogram
☐ square

5. Angles a and b combine to form a right angle. The angle measure of a is 40° less than the angle measure of b.

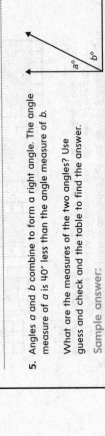

What are the measures of the two angles? Use guess and check and the table to find the answer.

Sample answer:

$a°$	$b°$	$a° + b°$	Is a 40° less than b?
10°	80°	90°	no, 70° less
40°	50°	90°	no, 10° less
35°	55°	90°	no, 20° less
20°	70°	90°	no, 50° less
25°	65°	90°	yes

$a =$ 25° , $b =$ 65°

Chapter 1 Test

NAME _____ DATE _____ SCORE _____

Chapter 1 Test

1. A scooter is priced between $1,000 and $2,000. Its price is a multiple of 10. All of the digits in the price, except for the thousands digit, are even numbers. The value of the hundreds digit is 30 times the value of the tens digit.

   Part A: Write the price of the scooter in standard form.

   $1,620

   Part B: Write the price of the scooter in word form.

   one thousand, six hundred twenty

2. The number on Ruben's ticket has a value between the values on the third and fourth raffle ticket shown.

   | 739075 | 738901 | 739214 | 740000 |

   A partial expansion of Ruben's ticket number is shown. Fill in the blanks to complete the expansion.

   7 × 100,000 + 3 × 10,000 + 9 × 1,000 + 9 × 100

3. In one month, a store sold between $3,000,000 and $4,000,000 worth of groceries. The estimate for the total sales that month had a 3 in the millions place, a 2 in the hundred thousands place, a 6 in the thousands place, and zeros in all of the other places.

   Part A: To what place was the total sales number most likely rounded? Explain your reasoning.

   It was probably rounded to the nearest thousand because the estimate has only zeros to the right of the thousands place.

   Part B: What would be the estimate if the total sales were rounded to the nearest ten thousand?

   $3,210,000

4. The numbers of runners who participated in three long distance races are shown on the table. The number of runners in the Bolder Boulder Run was less the number in the Great North Run but more than the number in the New York City Marathon. Select all possible numbers of runners in the Bolder Boulder Run.

Race	Number of Runners
Peachtree Road Race	55,500
Great North Run	54,500
New York City Marathon	50,740

   ☑ 54,040 ☐ 55,000
   ☐ 50,697 ☑ 51,681
   ☐ 54,600 ☐ 50,089

5. Use some of the base-ten blocks to model 1,323 in two ways.

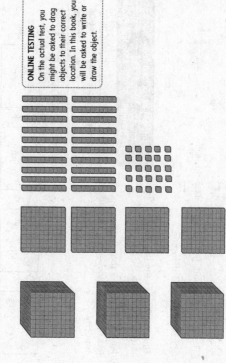

ONLINE TESTING On the actual test, you might be asked to drag objects to their correct location. In this book, you will be asked to write or draw the object.

1 thousand, 3 hundreds, 2 tens, 3 ones

1 thousand, 3 hundreds, 23 ones

Chapter 1 Test

6. Look at the table.

   Part A: Order the states from greatest (1) to least (4) according to number of acres of wilderness land area.

State	Wilderness Land Area (Acres)	Order
California	14,944,697	1
Colorado	3,699,235	3
Idaho	4,522,506	2
Minnesota	819,121	4

   Part B: Explain how you used place value to order the states.

   California has the only whose wilderness acreage with a ten millions place, so it has the greatest acreage. Then I looked at the millions place. The 4 in the millions place for Idaho is greater than the 3 in the same place for Colorado, so I ranked Idaho second, followed by Colorado. The number of acres of wilderness for Minnesota does not have a millions place, so I ranked it fourth.

7. A number is between 1 million and 8 million. The digit in the thousands place of the number is moved to a different position in the number so that the number's total value is 10 times greater.

   Part A: To which position did the digit move?

   The ten thousands place.

   Part B: After the digit is moved, the number is rounded to the nearest million. The result is greater than the original number. List all possibilities for the digit in the hundred thousands place of the original number. Explain your response.

   5, 6, 7, 8, 9; after the original number is rounded to the nearest million, the result is greater, which means that the millions digit increased. For the millions digit to increase when a number is rounded, the number in the hundred thousands place has to be 5 or greater.

8. More than eight hundred fifty-four thousand people live in a city. Select all numbers that could be the population of the city.

   - ☑ 8 × 100,000 + 55 × 1,000
   - ☑ 8 × 100,000 + 6 × 10,000 + 1 × 1,000
   - ☐ 807,230
   - ☐ 98,727
   - ☐ eight thousand, five hundred sixty-three
   - ☑ nine hundred thousand

9. The table shows the capacities of four football stadiums.

Professional Football Team Stadiums	
Team	Stadium Capacity
Atlanta Falcons	71,228
Baltimore Ravens	71,008
Houston Texans	71,054
San Diego Chargers	70,561

   Part A: Write the team names from least to greatest according to stadium capacity.

   | San Diego Chargers | < | Baltimore Ravens | < | Houston Texans | < | Atlanta Falcons |

   Part B: Which team's capacity has a digit with a value 10 times the value of the tens place digit in the Houston Texan's stadium capacity?

   San Diego Chargers

10. Three students round the same number to different places. Hector rounds to the thousands place and gets 15,000. Seiko rounds to the hundreds place and gets 14,700. Senthil rounds to the tens place and gets 14,650. What is the largest number that can be rounded as described?

    14,654

Chapter 1 Test

11. The numbers of several types of apples harvested on a farm are shown. Select **Right** or **Left** to describe the placement of the indicated number on a number line.

Type of Apple	Number Sold
Gala	316,489
Fuji	309,500
Macintosh	312,740
Honeycrisp	297,081

Left Right
☐ ☐ The number of Macintosh apples is to the _____ of the number of Fuji.
☐ ☐ The number of Honeycrisp apples is to the _____ of 300,000.
☐ ☐ The number of Gala apples is to the _____ the number of Macintosh.
☐ ☐ The number of Macintosh apples is to the _____ of 300,000.
☐ ☐ The number of Fuji apples is to the _____ the number of Gala.
☐ ☐ The number of Gala apples is to the _____ of 320,000.

12. Jessica is playing a game. The goal of the game is to use 7 number cubes to create the greatest possible number. Jessica rolls the numbers shown.

[number cubes: 0, 8, 7, 3, 9, 1, 2]

Part A: What is the greatest number that Jessica could create by rearranging the number cubes?

9,873,210

Part B: Round the number to the greatest possible place value.

10,000,000

13. Mrs. Norman's class recycled 3,412 pounds of newspaper. Mr. Rivera's class recycled 4,821 pounds of newspaper. Identify each digit whose value is 10 times greater in Mr. Rivera's number than in Mrs. Norman's number.

4 and 2

14. Jerome's city has a population of 1,567,231. Nhu's city has a population of 621,317. Select whether each statement is true or false.

True False
☐ ☐ The 6 in Nhu's city population has 10 times the value of the same digit in Jerome's.
☐ ☐ The leftmost 1 in Jerome's city population has 10 times the value of the leftmost 1 in Nhu's.
☐ ☐ The 2 in Nhu's city population has 10 times the value of the same digit in Jerome's.
☐ ☐ The 3 in Jerome's city population has 10 times the value of the same digit in Nhu's.
☐ ☐ The cities' populations would have the same value in the hundred thousands place if they were rounded to the nearest hundred thousand.

15. In the number shown, the digit in the ten thousands place is 4 times the digit in the tens place. The digit in the tens place is an even number and is not equal to 0. The value of the digit in the thousands place has 10 times the value of the digit in the hundreds place. What is the complete number written in word form?

5 ☐ ☐ , 1 ☐ 3

five hundred eighty-one thousand, one hundred twenty-three

Chapter 2 Test

NAME _____ DATE _____ SCORE _____

Chapter 2 Test

1. Use the digits 1, 2, 3, 4, 5, and 6 exactly once to create a subtraction problem between two 3-digit numbers with the greatest possible difference. Then round the difference to the nearest hundred.

   654 − 123 = 531; rounded: 500

2. The table shows the number of laptops sold by Computer World each month during the second half of the year. Select all pairs of months that have sales numbers which add to an estimate of 60,000 laptops.

   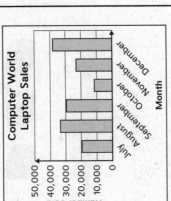
   Computer World Laptop Sales

   ☐ July and August
   ☒ July and December
   ☒ August and September
   ☐ August and November
   ☐ October and December
   ☒ November and December

3. David plans to run for 150 minutes every week. He ran for 28 minutes on Monday, 43 minutes on Tuesday, 39 minutes on Thursday, and 18 minutes on Friday. He estimates that he has about 20 minutes more to run to meet his goal. Write the correct numbers in the boxes to make an equation that shows how David could have come up with his estimate.

   | 10 | 20 | 30 | 40 | 50 | 150 |

   150 − (30 + 40 + 40 + 20) = 20

4. Nina is climbing to the top of Mt. Kilimanjaro. The mountain is 19,341 feet tall. Her base camp is at 9,927 feet. The guide for Nina's group has set up checkpoints at every 1,000-foot gain in elevation after the base camp.

   Part A: Complete the table below to show the checkpoint elevations.

	Elevation (ft)
First Checkpoint	10,927
	11,927
	12,927
	13,927
	14,927
	15,927
	16,927
	17,927
Last Checkpoint	18,927

   Part B: The guide also told the climbers to take a short rest break at every 100-foot gain in elevation past the last checkpoint. How many times should each hiker rest *after* the last checkpoint before reaching the top of the mountain? Justify your response.

   4 times; they should rest at 19,027 feet, 19,127 feet, 19,227 feet, and 19,327 feet. Those are the 100-foot intervals past 18,927 feet.

5. The table shows the year several states were admitted to the United States and the year that the states' first censuses were recorded. Sort the states by the time between the first census and the year admitted. Write the state names in the correct column of the table below.

State	Year Admitted to United States	First Census
Kentucky	1796	1790
Arkansas	1836	1810
Michigan	1837	1790
Texas	1845	1820
West Virginia	1863	1860
Colorado	1876	1860
Utah	1850	1896

   Number of Years Between First Census and Year Admitted to United States

Fewer than 10	Between 10 and 20	Between 20 and 30	More than 30
Kentucky	Colorado	Arkansas	Michigan
West Virginia		Texas	Utah

Chapter 2 Test

6. Three backpackers are sorting their shared gear so that weight of each person's backpack is within a certain range. Each backpack's maximum weight range and current weight is shown. There are four items remaining. Sort the items into the three backpacks so that their weights are within the given range.

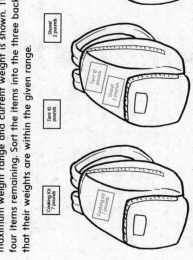

Backpack 1
Maximum Weight:
From 15 to 20 pounds
Current Weight:
12 pounds

Backpack 2
Maximum Weight:
From 20 to 25 pounds
Current Weight:
13 pounds

Backpack 3
Maximum Weight:
From 25 to 30 pounds
Current Weight:
25 pounds

7. A scientist recorded estimates of the number of bacteria in 4 colonies at the end of two 1-hour periods. Which colony had the greatest population growth, and by how many bacteria did it grow? Show your work.

Colony #	Hour 1	Hour 2
1	2,891,000	4,907,000
2	2,440,000	4,506,000
3	2,371,000	4,960,000
4	2,482,000	4,201,000

Colony 3; it grew by 2,589,000 bacteria. 4,960,000 − 2,371,000 = 2,589,000

8. The number 38,☐2 rounded to the hundreds place is 38,700. What is the greatest possible sum of the two missing digits? Justify your answer.

15; The hundreds digit could be 6 or 7. If it is 6, the tens digit could be 5, 6, 7, 8, or 9, giving sums of 11, 12, 13, 14, or 15. If it is 7, the tens digit could be 0, 1, 2, 3, or 4, giving sums of 7, 8, 9, 10, or 11.

9. The owner's manual for Shanese's car recommends that she get her car's oil changed every 3,000 miles. Last month, Shanese had her car's oil changed and then drove 1,074 miles. She drove 883 miles this month.

Part A: Write an equation with a letter standing for the unknown that can be used to find the number of miles Shanese can drive her car before it needs another oil change.

$3,000 - (1,074 + 883) = d$

Part B: Solve the equation. Show your work.

$3,000 - (1,074 + 883) = d$
$3,000 - 1,957 = d$
$1043 = d$

10. Ravi is saving money toward purchasing a bicycle that costs $350. He started this month with $78 in his savings account. He then saved $129 this month from his pay for delivering newspapers. His parents have agreed to pay $50 toward the cost of the bicycle. Select **all** expressions that are equal to the remaining amount that Ravi must save.

☐ 350 − 78 + 129 + 50
☐ 350 − (78 − 129) − 50
☒ 350 − (78 − 129 + 50)
☐ 350 − 78 − 129 − 50
☐ 350 − (78 + 129) + 50
☒ 350 − (78 + 129 + 50)
☐ 350 + 78 + 129 − 50

11. The table shows the current population of four cities and an estimate of the expected population increase or decrease over the next decade. Estimate each city's population, to the nearest thousand, at the end of the decade.

City	Population	Estimated Increase (+) or Decrease (−)
Belltown	31,897	+1,000
Claymore	121,043	−5,000
Monroe	19,796	−1,000
Stanton	63,662	+3,000

Belltown: 33,000
Claymore: 116,000
Monroe: 19,000
Stanton: 67,000

Chapter 2 Test

14. A house decreased in value from $200,000 to $189,500. The value then decreased again by $2,900. What was the total decrease in the value of the house? Show your work.

$13,400; 200,000 − 189,500 = 10,500; 10,500 + 2,900 = 13,400

15. The amount of tea left in four pitchers set out during a school meeting is shown at the right. The leftover tea was poured from all four pitchers into a 72-oz container. Shade the pitcher below to show the greatest amount of tea that can still fit into the 72-oz container.

16. The grid shows the seating chart for a school auditorium. The auditorium holds up to 225 people. The shaded cells show where adults are now seated to watch a play. In addition to the adults, there are currently 183 students in the auditorium. Select whether each statement is true (T) or false (F).

There are enough seats left for all of Mrs. Hill's 24 students.
If all of Mr. Lee's 25 students were seated in the auditorium, all of the seats would be filled.
There are enough seats left for all of Mr. Jackson's 28 students.
If 8 more adults are seated, there will not be enough seats for all of Mr. Dearwater's 20 students.
If 7 students leave their seats, there will be enough seats for all of Mrs. O'Dell's 32 students.

T	F			

64 Grade 4 · Chapter 2 Add and Subtract Whole Numbers

---

12. The table shows the number of cars that can park in each of the four lots at a mall. The mall's owner plans use the parking lot that holds the least number of cars to build a parking garage that will hold from 1,000 to 1,200 cars.

Mall Parking Lots	
Parking Lot	Capacity
A	795
B	492
C	642
D	847

Part A: The garage will be built to hold the greatest number of cars possible. What will be the greatest number of cars that will be able to park at the mall? Justify your response.

3,484; 795 + 642 + 847 + 1,200 = 3,484
I didn't add in parking lot B's number because the garage will be built on top of it.

Part B: The owner plans to build a second parking garage. What is the greatest number of cars that the second garage would need to hold so that a total of 4,500 cars can park at the mall? Justify your response.

1,216; 4,500 − (795 + 642 + 847 + 1,000) = 1,216
I used the least number that the first garage would hold.

13. Mr. Chavez took in $16,407 this month at his hardware store. He wrote a problem to subtract $9,154 he paid to his suppliers and $1,948 for expenses from what he took in. Shade in the box to select whether or not each statement describes a step that can be used to find how much money Mr. Chavez has left.

Yes	No	
		Use one hundred as 10 tens, which leaves 1 hundred and 15 tens.
		Subtract 4 tens from 4 tens to get 0 tens.
		Use one ten as 10 ones, which leaves 4 tens and 13 ones.
		Subtract 3 ones from 8 ones to get 5 ones.
		Use one thousand as 10 hundreds, which leaves 6 thousands and 12 ones.
		Subtract 9 hundreds from 12 hundreds to get 3 hundreds.

234 Grade 4 · Chapter 2 Add and Subtract Whole Numbers

NAME _____ DATE _____ SCORE _____

Chapter 3 Test

1. Beth has 15 seeds. She plants the same number of seeds in each of 3 flowerpots.

   Part A: Write a division statement whose answer is the number of seeds in each flowerpot. Include the answer in your statement.

   $15 \div 3 = 5$

   Part B: Represent the division statement on the number line.

   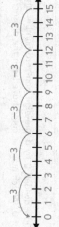

2. Select each statement that is a member of the fact family for the array.

   ☐ 42 − 6 = 36
   ☐ 42 − 36 = 6
   ▨ 42 ÷ 6 = 7
   ▨ 42 ÷ 7 = 6
   ☐ 6 × 6 = 36
   ▨ 6 × 7 = 42
   ▨ 7 × 6 = 42
   ☐ 7 + 6 = 13
   ☐ 6 + 7 = 13

3. A baker has 10 strawberries, 12 raspberries, and 10 blueberries. She plans to put the same total number of berries on each of 4 cakes.

   Part A: Before any calculations are done, is it reasonable to state that the baker will put fewer than 10 berries on each cake? Explain.

   Yes, because there are 3 types of berries, and around 10 of each type. So, if there were only 3 cakes, then each cake could have 10 berries. But since there are 4 cakes, each cake will have fewer than 10 berries.

   Part B: Write and solve an equation whose solution is *b*, the number of berries on each cake.

   $32 \div 4 = b$; $b = 8$

4. Seats are arranged in a theater as shown. Each circle represents a seat. Show two more arrangements that use the same number of seats.

   ONLINE TESTING
   On the actual test, you might be asked to drag objects to their correct location. In this book, you will be asked to write or draw the objects.

Chapter 3 Test

9. Two teams keep track of their points in a weekly math competition. For each week shown, indicate whether Team B has 4 times as many points as Team A.

Week	Team A	Team B	Does Team B have 4 times as many points as Team A?	
			Yes	No
1	||||	|||| ||||	✓	
2		|||		✓
3	|||| ||||	|||| |||| ||||		✓
4	|||| ||||	|||| |||		✓
5	||||	||||	✓	
6	|||| ||||	||||		✓
7	||||	|||		✓
8	|||| ||||	|||| ||||		✓

10. Marcus has 24 pencils, Melinda has 8 pencils, Jake has 5 pencils, and Suri has 32 pencils. For each pair of students, write a sentence comparing the number of pencils that they have.

Marcus and Melinda: Marcus has 3 times as many pencils as Melinda.

Melinda and Jake: Melinda has 3 more pencils than Jake.

Suri and Marcus: Suri has 8 more pencils than Marcus.

Suri and Melinda: Suri has 4 times as many pencils as Melinda.

---

5. Tanya divided cards from a game among her friends. Each friend received 13 cards. Tanya represented the division with the statement $52 \div \square = 13$.

   Part A: Write two multiplication facts for the division fact.

   $13 \times 4 = 52; \ 4 \times 13 = 52$

   Part B: How are the numbers in the multiplication facts related to the numbers in the division fact? Explain, using the appropriate vocabulary to identify each number.

   The factors in the multiplication facts, which are 4 and 13, are the divisor and quotient in the division fact.

6. Abby biked for 4 miles during the first week of summer vacation. During the last week, she biked 7 times as long as she did during the first week. Use numbers and symbols from the list shown to write an equation that can be used to find m, the number of miles that Abby biked during the last week of summer vacation.

   3   4   7   =   m   ×   ÷   +   −

   $m = 7 \times 4$

7. Dale divided 0 by a one-digit non-zero number and got a quotient of 0. Which number could have been the divisor? Explain.

   The divisor could have been any number from 1 to 9. Zero divided by any number equals zero.

8. Jay received 11 points on his first science project and 33 points on his second project. Write a statement that compares the number of points Jay received on the two projects. The solution to the statement should involve multiplication.

   Jay received 3 times as many points on his second project as he did on his first project.

Chapter 3 Test

11. Each table in the science lab has 3 containers of bugs that are identical to the one shown. Each container has 5 bugs. Select **all** number statements that can be used to determine the total number of bug legs at each table.

- ☐ 6 × 3 × 5
- ☐ 5 × (3 × 5)
- ☐ (3 × 5) × 6
- ☐ 30 × 3
- ☐ 5 × 3
- ☐ 1 × 3 × 5

12. A bookcase has 4 shelves. There are 9 books on each shelf.

Part A: Write two different multiplication sentences that show the total number of books in the bookcase.

4 × 9 = 36; 9 × 4 = 36

Part B: Write a third sentence that shows that the first two sentences are equal. Name the property that you used.

4 × 9 = 9 × 4; Commutative Property of Multiplication

13. Eun earns 3 stickers for each news article that she reads. She earned 27 stickers for reading *n* articles. Select whether each statement is true or false.

True	False	
☐	☐	The equation 3 × 27 = *n* can be used to find the number of articles that Eun read.
☐	☐	The equation 3 × *n* = 27 can be used to find the number of articles that Eun read.
☐	☐	The equation 27 ÷ 3 = *n* can be used to find the number of articles that Eun read.
☐	☐	Eun read exactly 81 articles.
☐	☐	Eun read exactly 9 articles.
☐	☐	Eun read exactly 3 articles.

14. Sidney, Darius, Jorge, Karla, and Hasan each picked the same number of tomatoes from the school garden. The total number of tomatoes is shown. Circle groups of tomatoes to model the number that each student picked.

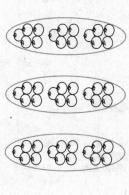

15. Kai found the product 8 × 3 × 1. First, he multiplied the 3 and the 1. Next, he multiplied that result by 8. Indicate whether each statement is true or false.

True	False	
☐	☐	Kai used the Commutative Property of Multiplication.
☐	☐	Kai used the Associative Property of Multiplication.
☐	☐	Kai used the Identity Property of Multiplication.
☐	☐	The result of the first operation that Kai performed is 24.
☐	☐	The final product is 24.

16. Miguel multiplied 0 by a one-digit non-zero number and got a product of 0. Which number could have been the other factor? Explain. Name the property that justifies this result.

The factor could have been any number from 1 to 9. Zero multiplied by any number equals zero. This is the Zero Property of Multiplication.

NAME _____ DATE _____ SCORE _____

Chapter 4 Test

1. Estimate each product by rounding the larger number to the greatest place value possible. Then check the appropriate box to indicate whether the estimate is greater than or less than the actual product.

Product	Estimate	Estimate is Less than Actual Product	Estimate is Greater than Actual Product
119 × 7	700	✓	
2,801 × 9	27,000		
169,032 × 4	800,000		✓
23,698 × 4	80,000	✓	

2. Look at the base-ten blocks.

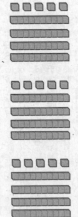

Part A: What product do the base-ten blocks represent?

45 × 3

Part B: Explain how to use the base-ten blocks and place value to find the product. Describe any regrouping that you must do.

There are 3 × 4 = 12 tens. Regroup 10 tens as 1 hundred, which leaves 2 tens. There are 3 × 5 = 15 ones. Regroup 10 ones as 1 ten. That gives me a total of 3 tens and 5 ones. So the number has 1 hundred, 3 tens, and 5 ones. The number is 135.

Chapter 4 Test

ONLINE TESTING On the actual test, you might use a keyboard to type in answers. On this test, you will write the answers.

3. The Ross family pays $105 each month for their cable and internet. Write the correct numbers in the expression to show the total amount, in dollars, that the Ross family pays for cable and internet in 6 months.

(6 × [100]) + (6 × [0]) + (6 × [5])

4. As an assignment for art class, a group of art students completed information forms on different paintings. They visited 3 museums and completed 26 forms at each museum.

Part A: Draw and label the area model to represent the total number of forms the students completed. Show the partial areas.

Part B: How many forms did the students complete? Use your area model to explain how you know.

78; 26 × 3 = 60 + 18 = 78

5. Four classes are going to see a play. Each class has 23 students.

Part A: Use base-ten blocks to model the total number of student tickets needed by the four classes.

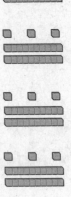

Part B: How many tickets are needed? Explain how your model shows this.

92; I started with 2 tens rods and 3 ones units to get 23. I made 3 more copies of this to get a total of 4 copies to represent the 4 classes. This gives a total of 2 × 4 = 8 tens rods, which is 80, and 3 × 4 = 12 ones units, which is 12. 80 + 12 = 92.

Chapter 4 Test

6. Select **each** expression that represents the model shown.

   ☐ (4 × 10) + (8 × 10)
   ☐ 4 × 10 × 8
   ☒ (4 × 10) + (4 × 8)
   ☐ 16 × 18
   ☒ (4 × 10) × (4 × 8)
   ☐ 4 × 18

7. Look at the area model.

   ```
        40        2
     ┌────────┬──────┐
   3 │ 40×3=120│2×3=6│
     └────────┴──────┘
   ```

   Part A: Write the partial products represented by the area model. Find the sum of the partial products.

   40 × 3 = 120
   2 × 3 = 6
   120 + 6 = 126

   Part B: Write the complete number sentence that is represented by the area model.

   42 × 3 = 126

8. Lila estimated the product ☐ 352 × 6 as 30,000 by first rounding to the greatest place value possible. What is the missing digit?

   5

9. The tripmeter shows the number of miles that a delivery driver travels on one of his routes. He travels the route once each day on Monday, Tuesday, Wednesday, and Thursday. Select **each** statement that correctly describes the estimate of the number of miles the driver travels on the route each week.

   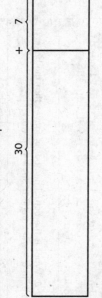

   ☐ The estimate is less than the actual number of miles.
   ☒ The estimate is greater than the actual number of miles.
   ☐ The product 300 × 4 can be used to find the estimate.
   ☐ The product 300 × 7 can be used to find the estimate.
   ☒ The product 400 × 4 can be used to find the estimate.
   ☐ The product 400 × 7 can be used to find the estimate.

10. Dillon used the area model shown to represent 37 × 7.

    ```
           30       +  7
         ┌──────────┬──┐
       7 │          │  │
         └──────────┴──┘
    ```

    30 × 7 = 210
    210 + 7 = 217

    Part A: He used partial products as shown to calculate the product 37 × 7. What mistake did Dillon make?

    He did not multiply 7 × 7 before adding it to 210.

    Part B: Use partial products to find the actual product. Show your work.

    30 × 7 = 210
    7 × 7 = 49
    210 + 49 = 259

Chapter 4 Test

11. You want to multiply the numbers shown as they are written. For each product, select whether you would have to use regrouping.

    ONLINE TESTING
    On the actual test, you might click to select options. In this book, you will use your pencil to shade the correct boxes.

Requires Regrouping?		
Yes	No	
☐	☐	2,08 × 5
☐	■	1,133 × 2
☐	☐	4,509 × 3

12. As part of a fundraiser, each grade level has a goal of raising $2,250. How much will the 6 grade levels raise if they all meet their goals? Show your work.

    1 3
    2,250
    × 6
    ―――
    13,500

    They will raise $13,500.

13. An assembly line in a factory fills 1,130 bottles of juice every hour. Indicate whether each statement is true or false.

True	False	
☐	☐	It is reasonable to state that the assembly line will meet its goal of filling 10,000 bottles during one 8-hour work day.
☐	☐	A good estimate for the number of bottles filled during a 3-hour time period is 3,000.
■	☐	A supervisor would have to do some regrouping to find the number of bottles that the assembly line can fill in 5 hours.
■	☐	The assembly line can fill twice as many bottles in 4 hours as it can fill in 2 hours.
☐	■	The assembly line can fill at most 6,680 bottles during a 6-hour time period.

14. Use your knowledge of place value and multiplication to find the missing digits in the product shown. Write the missing digits in the boxes.

    1,6[7]3 × [4] = 6,692

15. A bakery makes 2,094 loaves of bread each week. The manager calculated that the bakery produces 8,076 loaves in a 4-week period.

    Part A: What mistake did the manager make in calculating the total? Use place value in your explanation.

    When multiplying 9 in the tens place by 4, you get 36, plus 1 more ten regrouped from the ones place, which makes 37 tens. You write down 7 tens and then regroup 30 tens as 3 hundreds. The manager did not regroup the 3 hundreds or forgot to add them to the 0 hundreds that result from the product of 0 and 4.

    Part B: What is the actual number of loaves that the bakery makes in a 4-week period? Show your work.

    3 1
    2,094
    × 4
    ―――
    8,376

    The bakery makes 8,376 loaves.

16. A local farm cooperative charges $98 per month for its family produce plan. Select **each** expression that can be used to find the amount a family must pay for a 6-month plan.

    ☐ (6 × 9) + (8 × 9)
    ■ 98 × 6
    ☐ (6 × 90) + (6 × 8)
    ☐ (6 × 90) × (8 × 90)
    ■ 6 × 8 + 6 × 90
    ☐ 90 × 8 + 6

Chapter 5 Test

NAME _____ DATE _____ SCORE _____

Chapter 5 Test

ONLINE TESTING
On the actual test, you might click to select options. In this book, you will use a pencil to shade the correct boxes.

1. Mr. Gibson passed out the same number of cubes to each student in his class.
   - The number of cubes each student has is greater than 20.
   - Each student has an odd number of cubes.
   - Mr. Gibson passed out a total of 540 cubes.

   Select all statements that are true.
   - ☐ Mr. Gibson could have 54 students in his class.
   - ☒ Mr. Gibson could have 40 students in his class.
   - ☒ Mr. Gibson has an even number of students in his class.
   - ☐ The number of students in Mr. Gibson's class is less than 30.
   - ☒ If Mr. Gibson had passed out 1080 total cubes, each student would have twice as many cubes.

2. Estimate each product by rounding each number to the greatest place value possible. Then check the box to indicate whether the estimate is less than or greater than the actual product.

Product	Estimate	Estimate is Less than Actual Product	Estimate is Greater than Actual Product
21 × 53	1,000	✓	
89 × 64	5,400	✓	
55 × 31	1,800		✓
72 × 40	2,800	✓	

3. Which number sentence can you use as a first step to help you find 21 × 70?

   A. 21 ÷ 7 = 28
   B. 21 + 70 = 91
   C. 21 × 7 = 147
   D. 21 ÷ 3 = 7

---

4. A theater has 24 rows. Each row has 16 seats.

   Part A: Which expression shows the total number of seats in the theater?

   A. (20 × 10) + (4 × 6)
   B. (24 × 10) + (24 × 6)
   C. (20 × 6) + (4 × 10)
   D. (24 × 6) × (24 × 10)

   Part B: Complete the area model. Select from the numbers shown. Then write the total number of seats in the theater. The model should match the expression from **Part A**.

   Numbers: 4 6 10 16 20 24 40 144 200 240

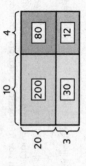

   Number of Seats = 384

ONLINE TESTING
On the actual test, you might drag and drop numbers into the boxes. In this book, you will write the numbers in the boxes.

5. Ms. Gonzalez passed out some worksheets to her class. The total number that she passed out is shown by the area model.

	20	3
10	200	30
4	80	12

   Part A: Write an equation that shows the product that the area model represents.

   23 × 14 = 322

   Part B: Complete the statement below. Notice that there are two different ways that it could be completed.

   Ms. Gonzalez passed out 14 worksheets to each of the 23 students in her class.

Chapter 5 Test

6. Adele has 12 solid-colored marbles and 27 multicolored marbles in her collection. Wei has 13 times as many marbles in his collection. Select all equations that can be used to find m, the number of marbles in Wei's collection.

   ☐ $39 \div m = 13$
   ☐ $m = (12 \times 13) + (27 \times 13)$
   ☐ $m \div 13 = 39$
   ☐ $m = (12 \times 10) + (27 \times 3)$
   ☐ $m = 13 \times 39$
   ☐ $39 \div 13 = m$

7. A student made an error in the multiplication problem shown.

   ```
       65
    ×  42
    ─────
      130
    +260
    ─────
      390
   ```

   Part A: Explain the student's error, using place value.

   Sample answer: When multiplying by tens in the second row, the student forgot to put a 0 in the ones place to show 0 ones.

   Part B: Write a number in each of the boxes to correctly complete the problem.

   ```
         65
      ×  42
      ─────
      1 3 0
   +2 6 0 0
   ─────────
    2 7 3 0
   ```

8. Which number correctly completes the equation?

   $34 \times 87 = 2400 + 210 + \boxed{} + 28$

   A. 32
   B. 320
   C. 272
   D. 2720

9. A parking lot has 15 rows. Each row has p parking spaces. The number p is an even 2-digit number. Indicate whether each statement is true or false.

	True	False
The lot has an even total number of parking spaces.	☐	☐
The total number of parking spaces could be less than 100.	☐	☐
The number p must have a zero in the one place.	☐	☐
The total number of parking spaces is a multiple of 10.	☐	☐
The number equal to $15 \times p$ could be greater than 1,000.	☐	☐

10. The pictograph shows the number of cans collected by students at Riverview Elementary during a recycling drive. Write the correct number in each box to make an equation that models the situation.

    Number of Cans Collected

    Grade 1: 🥫🥫🥫🥫
    Grade 2: 🥫🥫🥫🥫🥫
    Grade 3: 🥫🥫🥫
    Grade 4: 🥫🥫🥫🥫🥫
    Grade 5: 🥫🥫
    Grade 6: 🥫🥫🥫

    key: 🥫 = 25 Cans

    $\boxed{19} \times \boxed{25} = \boxed{475}$

11. Write the missing digits that complete the multiplication problem.

    ```
          3 9
       ×  4 3
       ──────
        1⎕ 7
      +15 6 0
      ───────
       16 7 7
    ```

Chapter 5 Test

12. A recreation center orders 38 cartons of tennis balls. Each carton contains 6 of the containers shown in the picture.

Part A: Write the correct numbers and symbols in the boxes to make an equation that shows an estimate of the number of tennis balls ordered. Select from the list below.

10 20 30 **40** 200 300 400 600 **800** 1,200

+   −   **×**   ÷

[40] [×] [20] = [800]

Part B: Is the estimated number of tennis balls ordered greater than or less than the actual number? Explain how you know.

Sample answer: The estimate is greater. There are 6 × 3 = 18 balls per carton, which I rounded up to 20. There are 38 cartons, which I rounded to 40. Because the estimate numbers are greater than the actual numbers, the estimated product is also greater.

13. Mr. Holt has 26 students in his class. He passed a bin of markers around the room 11 times. The first 10 times that the bin came around, every student selected a marker. The last time, only some of the students selected a marker. Others did not get to select a marker because the bin was empty before it got to them. Select **all** choices that could be the original number of markers in the bin.

☐ 26   ☐ 261   ■ 280
☐ 37   ■ 260   ☐ 286
☐ 47   ■ 272

14. In a cycling relay race, each team member rides for 16 kilometers. Team Rowdy Riders has 11 team members. On race day, all of the Rowdy Riders carpooled to the race. They rode in the race, and then rode their bikes to a lake to cool off. Finally, they rode back to the race site. The lake is 3 kilometers from the race site. For what total distance did the Rowdy Riders ride their bikes that day?

A. 143 kilometers   C. 209 kilometers
B. 176 kilometers   (D.) 242 kilometers

15. Corny's Roasted Corn stand bought 35 dozen ears of corn for the fair. Corny's sold 130 ears on the first day of the fair, and 162 on the second day. Corny's sold the rest of the ears on the third day. Indicate whether each statement is true or false.

True	False	
☐	☐	More ears of corn were sold on the third day of the fair than on the first or second day.
■	☐	Corny's started the fair with more than 400 ears of corn.
■	☐	The equation 35 × 12 − 130 + 162 = c can be used to figure out the number of ears of corn sold on the third day.
☐	☐	The expression 35 × 12 can be used to figure out the total number of ears of corn sold at the fair.

16. A group of 74 students and 8 teachers order tickets for a theater festival. Student tickets cost $12 each, and teacher tickets cost $15 each. There is a fee of $5 for each group order.

Part A: Complete the equation so that it shows the total amount, t, that the group paid for the order.

74 × [12] + [8] × [15] + [5] = t

Part B: Solve the equation. Show your work and state the answer.

74 × 12 + 8 × 15 + 5 = t
888 + 120 + 5 = t
888 + 125 = t
1,013 = t

The order cost $1,013.

Chapter 6 Test

NAME _____ DATE _____ SCORE _____

Chapter 6 Test

1. A charity sends out 9 volunteers to conduct a survey of 270 households. Each volunteer will survey the same number of households. Explain how to use a basic fact and place value to find the number of households each volunteer will survey.

   Sample answer: Divide 27 by 9 to get 3. The answer is then 10 times more than 3 because 270 is 10 times more than 27. So, each volunteer will survey 30 households.

   ONLINE TESTING On the actual test, you might click to select options. In this book, you will use a pencil to shade the boxes.

2. Gianna is going on a cross-country trip with her family. They plan to drive 2,046 miles over 7 days. Gianna estimates that her family will drive about 300 miles per day. Select **each** true statement about Gianna's calculation.

   ☐ Gianna rounded to the nearest hundred before dividing.
   ☐ Gianna used compatible numbers to estimate.
   ☐ Gianna's estimation can be checked with the product 300 × 7.
   ☐ Gianna's estimation can be checked by adding 46 to 2,000.
   ☐ Gianna's estimate is greater than the actual distance the family will drive if they travel the same number of miles each day.

3. The cafeteria manager must divide 63 bagels equally among 3 different fourth grade classes. The model below shows how he did this.

   Part A: Complete the division sentence shown by the model. $\boxed{63} \div \boxed{3} = \boxed{21}$

   Part B: What does the answer to the division sentence mean in terms of the problem situation?

   The manager gave each class 21 bagels.

4. Ms. Simmons spent a total of $520 at the animal clinic for her dog and three cats. She spent the same amount on each pet.

   Part A: Model the total amount Ms. Simmons spent using the least number of base-ten blocks possible.

   Part B: Show how to use the base-ten blocks to solve the problem. You may trade for different base-ten blocks, if necessary.

5. Mark "Y" for **each** division statement that has a remainder and "N" for **each** that does not.

Y	N	
☐	☐	132 ÷ 10
☐	☐	65 ÷ 5
☐	☐	231 ÷ 8
☐	☐	56 ÷ 7
☐	☐	136 ÷ 4

Chapter 6 Test

ONLINE TESTING
On the actual test, you might use a keyboard to type in answers. In this book, you will write the answers.

8. Write the missing digits that complete the division problem.

    ☐)4R3
    6)13☐7

9. A company has 593 large and 767 small packages. The company has 20 delivery trucks. Each truck will deliver the same number of packages to a shipping port. Use some of the numbers and symbols shown to write an equation that can be used to find p, the number of packages that each truck will carry.

    (   )   +   −   ÷   ×   =   593   767   20 p

    $(593 + 767) \div 20 = p$

10. Jennae plans to buy one of the computers whose price is shown. She wants to divide the cost of the computer into 4 payments. Write the prices in the correct bin according to whether each computer can be paid for with 4 payments of the same exact whole dollar amount.

4 Equal Whole Dollar Payments	4 Unequal Whole Dollar Payments
$1,104	$437
$784	$254
	$1,310

11. Jon has 533 pennies. He wants to put the same number of coins in each of 6 piggy banks. He uses the division statement 540 ÷ 6 to estimate that he will put about 90 pennies in each bank. Ellie states that Jon estimated wrong because 533 does not round to 540. Is Ellie correct? Explain.

    Sample answer: Ellie is correct in stating that 533 does not round to 540, since it rounds to 530. But, Jon did not estimate wrong because he used compatible numbers. 530 does not divide evenly by 9, but 540 does, so that is what Jon used.

---

6. Shawn finds 5)1,545 and gets 39. He checks it and gets 39 × 5 = 195. State which calculation was wrong: the division or the multiplication. Then use place value to explain the mistake that Shawn made.

    Sample answer: His division is wrong. 1 in the thousands place does not divide by 5, so he starts with 15 hundreds divided by 5 to get 3 hundreds. 3 times 5 is 15, so there are no hundreds left. Then he divides 4 tens by 5, which cannot be done, so he needs to put a 0 above the 4 to show that there are 0 tens in the answer. He forgot to put the 0 above the 4.

7. The fourth graders at Mountainview Elementary are sorting canned food donations into boxes for a community kitchen. The table shows the number of each type of donation.

Type of Food	Number of Cans
Peaches	32
Pears	45
Peanut butter	37
Black or pinto beans	66

    Part A: How many boxes can be packed if each box must contain 8 cans that contain a protein food (beans or peanut butter)? Show your work, and tell how you know whether there will be any protein food cans left over.

    66 + 37 = 103. 103 ÷ 8 = 12 R7; 12 boxes can be packed. There is a remainder, 7, which means that there are protein food cans left over.

    Part B: The cans of fruit will be divided among the number of boxes determined in **Part A**. Explain how to find the number of cans of fruit that will go into each box.

    There are 32 + 45 = 77 total cans of fruit to go into 12 boxes. 77 ÷ 12 = 6 R5, so each box will get at least 6 cans of fruit.

Chapter 6 Test

12. Four third grade classes, 3 fourth grade classes, and 3 fifth grade classes are selling wrapping paper to raise money. Each class will sell the same number of rolls of each type of wrapping paper. The numbers of each type are shown in the table.

Wrapping Paper Fundraiser

Type of Paper	Number of Rolls	Rolls per Class	Rolls Left
Valentine's Day	37	3	7
Birthday	223	22	3
Baby Shower	50	5	0
Wedding	185	18	5
Winter Holiday	360	36	0

Part A: How can you tell, without dividing, whether the number of rolls of each type can be divided evenly among the classes?

Sample answer: $4 + 3 + 3 = 10$ classes will sell the paper. If a number is divisible by 10, then it ends in a zero. So, only the papers with a count that ends in zero can be evenly divided.

Part B: Find how many rolls of each type each class will sell. State how many rolls of each type, if any, will be left over. Write the values in the table.

13. Mr. Jeong wants to figure out how much money he can spend on his garden center expenses for each of the first 7 weeks after it opens. He uses the problem $7\overline{)4,340}$. Determine whether each statement is true or false.

True False
◯ ◯ His problem has a remainder.
◼ ◯ He can spend $620 each week.
◯ ◼ He can spend $627 each week.
◼ ◯ He starts with a total of $4,340 for expenses.
◯ ◼ He starts with a total of $30,380 for expenses.

14. Sani is in a group of 4 art students. He wants to figure out how many sheets of construction paper each student in his group should get. He has 234 sheets. Sani will check his work. Write the numbers 1–3 next to the correct three steps below to show the order in which Sani should perform them.

    3 Add 2 to 232.
    1 Divide 234 by 4.
    __ Divide 58 by 4.
    2 Multiply 58 by 4.
    __ Subtract 2 from 234.

15. A company orders 337 desks. Four of them are used for the reception area. The rest of the desks will go on the company's 3 floors of offices. Each floor will get the same number of desks.

Part A: Write an equation that can be used to find d, the number of desks that will go on each floor.

$333 \div 3 = d$

Part B: Think about the numbers in the division problem. Explain how you can solve the problem using place value and mental math. Then state the answer.

Sample answer: I have 3 hundreds, 3 tens, and 3 ones. If I divide each by 3, I get 1 hundred, 1 ten, and 1 one. So, there will be 111 desks on each floor.

16. Eduardo has a school supply allowance of $50. He bought 2 packs of pencils for $2 each, a pencil sharpener for $1, a pack of markers for $9, and a backpack for $25.

Part A: How many $3.00 notebooks can Eduardo buy with the remaining money? Show your work.

$2 + 2 + 1 + 25 + 9 = 39$; $50 - 39 = 11$; $11 \div 3 = 3$ R2; He can buy 3 notebooks.

Part B: Will Eduardo have any money left over? If so, how much? Explain how you know.

Yes, there is a remainder of 2, so he will have $2.00 left over.

Chapter 7 Test

NAME _____ DATE _____

SCORE _____

Chapter 7 Test

1. Look at the number list.

   Part A: Extend the pattern.

   48, 46, 51, 49, 54, 52, __57__, __55__

   Part B: Complete the statement.

   The pattern is ___subtract 2, add 5___.

2. Aaron is making a pattern with toothpicks. Each day, he adds to the pattern, as shown in the picture. Select all true statements below about the pattern.

   Day 1

   Day 2

   Day 3

   ☐ It is a growing pattern.
   ☐ A rule for the pattern is "Add one rectangle."
   ☐ It will have exactly 24 squares on the 8th day.
   ☐ It will have exactly 70 toothpicks on the 10th day.
   ☐ Aaron can figure out the number of squares it will have on the 12th day without actually continuing the pattern.

   ONLINE TESTING
   On the actual test, you might click to select options. In this book, you will use a pencil to shade the correct boxes.

3. Carrie is decorating the front of her notebook with stickers in the pattern shown.

   ♡ ☽ ☽ ☆ ♡ ☽

   Part A: Draw the next three stickers that Carrie will use.

   ☽ ☆ ♡

   Part B: Is this a growing pattern or a repeating pattern? Explain how you know.

   Sample answer: repeating; The heart, two moons, and star keep repeating in the same exact pattern. A growing pattern would have a greater number of figures each time.

4. The table shows the colors of flowers in several rows of a tulip farm.

Row	Color
1	red
2	purple
3	blue
4	purple
5	red
6	purple
7	blue
8	purple

   Part A: Complete the table. Then circle the rows necessary to show the pattern unit.

   Part B: Compare the number of rows of purple flowers to the number of rows of red and rows of blue flowers in one pattern unit.

   Sample answer: The number of rows of purple flowers is twice the number of either red or blue flowers.

5. Ms. Chopra listed the first 10 multiples of 8 on the board. Select all correct statements.

   ☐ All of the numbers on the list are even.
   ☐ The ones digits in the list repeated the following pattern: 8, 6, 4, 2, 0
   ☐ The sum of the digits in each number in the list is divisible by 4.
   ☐ Each number in the list is divisible by 4.
   ☐ Each number in the list is divisible by 16.

Grade 4 · Chapter 7 Patterns and Sequences

Chapter 7 Test

6. Lena picks 3 tomatoes from her garden on Monday, 3 on Tuesday, 6 on Wednesday, 12 on Thursday, and 24 on Friday. She notices that the number she picks each day after Monday is the sum of the numbers she picked on all previous days of the week. What is another pattern that describes the number picked on any day after Monday?

   After the first day, the number she picked on any day is twice the number she picked the day before.

7. Each day for 5 days, Tolu uses 9 sheets of notebook paper. Which pattern could be a list of the number of sheets he has left at the end of each day?

   A. 9, 18, 27, 36, 45
   B. 80, 71, 63, 54, 45
   C. 10, 19, 28, 37, 46
   D. 65, 56, 47, 38, 29

8. Kimiko is training for a long distance race. She sets a goal to run 5 kilometers the first week. During the first week, she runs 6 kilometers farther than her goal. Each week, she runs 6 kilometers farther than she ran the previous week.

   Part A: How far does she run during the 8th week?

   53 kilometers

   Part B: Complete the equation with the values shown so that it can be used to find the distance, k, that Kimiko runs during week w.

   5   6   k   w

   $k = \boxed{6} \times \boxed{w} + \boxed{5}$

9. The manager of a dinner theater uses the equation shown to figure out the amount of money, m, the theater earns when p people order a dinner and a performance. What is the value of m when $p = 18$? Show your work.

   $(8 + 12) \times p = m$

   $(8 + 12) \times p = m$
   $20 \times 18 = m$
   $m = \$360$

10. Look at the table.

Input (r)	4	8	12	16
Output (v)	12	24	36	48

    Which equation describes the pattern?

    A. $r \times 3 = v$
    B. $v \times 3 = r$
    C. $r + 8 = v$
    D. $v + 8 = r$

11. Use order of operations to determine the unknown that completes the equation.

    $(7 + 4) \times (6 - \boxed{2}) = 44$

12. Tim is finding output values for the equation $100 \div (a + 1) = t$. He states that the output, t, is 26 when the input, a, is 4.

    Part A: What is Tim's mistake?

    He did the operations in the wrong order. He divided 100 by 4 first. Because of the parentheses, he should have added 4 and 1 first, and then divided 100 by the result.

    Part B: Rewrite the equation so that Tim's value is correct.

    $(100 \div a) + 1 = t$

13. A pattern is generated using this rule: start with the number 81 as the first term and divide by 3. Enter numbers into the boxes to complete the table.

Term	Number
First	81
Second	27
Third	9
Fourth	3
Fifth	1

Chapter 7 Test

14. A new café sells 22 cups of tea on the first day of the week. It sells 30 on the second day, and 38 on the third day. This pattern continues for the rest of the week. Indicate whether each statement is true or false.

	True	False
The number of cups of tea sold by the café each day is always a multiple of 8.	☐	☐
The rule for the number of cups of tea sold the next day is "add 8".	☐	☐
The total number of cups of tea sold by the café during the week is even.	☐	☐
The increase in the number of cups of tea sold each day forms a pattern.	☐	☐

15. A store clerk arranges groups of oranges in the pattern shown. Draw the next group in the pattern.

16. Extend the pattern.

17. In the table shown, each rule's input is the previous rule's output.

Part A: Complete the table by writing in the missing values.

Input	Add 4	Subtract 2	Multiply by 3
5	9	7	21
6	10	8	24
10	14	12	36
3	7	5	15

Part B: If n is the input, write an equation that can be used to find q, the output.

$(n + 4 - 2) \times 3 = p$

18. The table shows the number of gallons of gas used by a truck driver on different trips and the amount paid for gas.

Input (gallons of gas)	24	203	89	327
Output (amount paid for gas)	$96	$812	$356	$1,308

Based on the table, how is the output obtained from the input?

A. Add 4.
B. Multiply by 4.
C. Add 72.
D. Multiply by 48.

19. How many rings will be in the 12th figure in the pattern shown?

23

Chapter 8 Test

NAME _____ DATE _____ SCORE _____

Chapter 8 Test

1. Roscoe arranged 80 square tiles in several rectangular patterns. Each pattern had more than 2 rows and had no tiles left over.

   Part A: Write the dimensions of the rectangular patterns Roscoe could use.

   Dimensions: 4 by 20, 5 by 16, 8 by 10

   Part B: Roscoe wants to make a square pattern out of his tiles. If he makes the largest square possible, what will the dimensions of the square be? How many tiles will be left over? Draw and label Roscoe's square.

   Dimensions of square: 8 by 8; 16 tiles left over

2. A tech website ranked four printers with respect to printing speed. The table shows the amount of time each printer took to print one page.

   Part A: Rank the printers from fastest to slowest on the table.

Printer	A	B	C	D	E
Time to print 1 page (minutes)	$\frac{1}{6}$	$\frac{3}{8}$	$\frac{5}{12}$	$\frac{1}{4}$	$\frac{5}{24}$
Rank	1	4	5	3	2

   Part B: Where would a printer that took $\frac{2}{5}$ of a minute to print a page rank on the table?

   Between printer B and printer C, slower than all but printer C.

3. The City Council wants to pass a law that bans the use of plastic shopping bags. The law states that at least $\frac{5}{8}$ of the council must vote yes for a law to pass. The Council has 40 members. Select true or false for each statement.

True	False	
☐	☐	The measure will pass with 24 yes votes.
☐	☐	The measure will pass if $\frac{3}{5}$ of the members vote yes.
☐	☐	The measure will pass with 25 yes votes.
☐	☐	The measure will pass if $\frac{15}{20}$ of the members vote yes.

4. Monica is cutting slats to make a butcher block. She cut 11 slats. The end of each slat is a square that is $\frac{1}{4}$ foot on each side.

   Part A: If Monica stacks the slats on their long sides, how tall will the butcher block be?

   $2\frac{3}{4}$ ft

   Part B: Draw and label Monica's butcher block.

5. A 1-mile horse race is measured in furlongs. There are exactly 8 furlongs in 1 mile, so each furlong is $\frac{1}{8}$ of a mile. If a horse has completed 6 furlongs in a 1-mile race, how far has it run? Select the correct distances.

   ☐ $\frac{3}{4}$ mile ☐ 6 miles
   ☐ 3 furlong ☐ $\frac{6}{8}$ mile

Chapter 8 Test

6. Jadamian claims that half of the odd numbers that are less than 100 are prime numbers. Is Jadamian correct? Justify your answer.

50 odd numbers: 1, 3, 5, 7, 9, 11, 13, 15, 17, 19, 21, 23, 25, 27, 29, 31, 33, 35, 37, 39, 41, 43, 45, 47, 49, 51, 53, 55, 57, 59, 61, 63, 65, 67, 69, 71, 73, 75, 77, 79, 81, 83, 85, 87, 89, 91, 93, 95, 97, 99

24 odd prime numbers: 3, 5, 7, 11, 13, 17, 19, 23, 29, 31, 37, 41, 43, 47, 53, 59, 61, 67, 71, 73, 79, 83, 89, 97. $\frac{24}{50} < \frac{1}{2}$, Jadamian is incorrect.

7. At Bear Lake, Ranger Dina takes care of reserving camp sites. So far $\frac{1}{2}$ of the sites have been reserved for next week.

ONLINE TESTING On the actual test, you might be asked to drag objects to their correct location. In this book, you will be asked to write or draw the object.

Part A: Dina has told all of the campers that their site does not share a side with another reserved site. Is Dina correct? Color in sites to justify your answer.

Yes, it is possible; the pattern shown has $\frac{1}{2}$ of the campsites reserved and no site is next to another.

Bear Lake Campsites

Part B: Three more campers have now reserved sites. How will that change Dina's statement that no one will have a site that shares a side with another reserved site? Color in sites to justify your answer.

Now at least 7 of the campers will have direct neighbors.

Bear Lake Campsites

8. Rajiv conducted a survey of students' attitudes toward texting and phone calls. The results are shown in the table.

Class	Prefer texting to phone	Total students
Grade 4	12	18
Grade 5	15	25
Grade 6	14	20
Grade 7	24	36

Part A: Which class had the greatest preference for texting over phone calls? Justify your answer.

The fraction of 6th graders was greatest, $\frac{7}{10}$ of the total. $\frac{7}{10}$ is greater than $\frac{2}{3}$ for Grade 4 and Grade 7, and $\frac{3}{5}$ for Grade 5.

Part B: The eighth grade data for 30 students showed that eighth graders had the same preference for texting as sixth graders. How many eighth graders preferred texting over phone calls?

21

9. Batting records of baseball players were compared.
   - Diaz was a better hitter than Williams.
   - Diaz was not as good of a hitter as Cabrera.

Each number shows the fraction of the time that each player got a hit. Which of the following records is possible for the three hitters?

☐ Diaz: $\frac{1}{3}$, Williams: $\frac{3}{10}$, Cabrera: $\frac{2}{5}$

☐ Diaz: $\frac{4}{10}$, Williams: $\frac{3}{10}$, Cabrera: $\frac{2}{5}$

☐ Diaz: $\frac{3}{8}$, Williams: $\frac{1}{3}$, Cabrera: $\frac{4}{9}$

☐ Diaz: $\frac{1}{4}$, Williams: $\frac{1}{3}$, Cabrera: $\frac{3}{8}$

Chapter 8 Test

10. Jana is building a wooden table and needs to measure a peg to fit into a hole. Jana knows that the hole is between $3\frac{3}{4}$ and $3\frac{9}{10}$ inches in width, but does not have a tape measure to find the exact width. Which of the following could be the actual width of the hole? Select all that apply.

    ☐ $3\frac{7}{10}$ inches   ☐ $3\frac{5}{6}$ inches
    ☐ $3\frac{4}{5}$ inches   ☐ $3\frac{11}{12}$ inches

11. Moncef, the owner of a software start-up, wants his 48 workers to split up into creative teams.

    - Each team must have more than 2 members.
    - Each team must have the same number of members.

    How many different arrangements of teams can the workers form? List them all in the box below.

2 teams of 24 members each	3 teams of 16 members each
4 teams of 12 members each	6 teams of 8 members each
8 teams of 6 members each	12 teams of 4 members each
16 teams of 3 members each	

12. Solve this riddle.

    Part A: I am a composite number between 50 and 60. Both of the factors that form me are prime numbers themselves. Which numbers might I be? Circle the possible numbers.

    50, (51), 52, 53, 54, (55), 56, (57), 58, 59, 60

    Part B: I am one of the numbers from Part A above. The sum of my factors equals a composite number that is a square of another number. Which number am I? Explain.

    55; the sum of 11 and 5 is 16, the square of 4.

13. The box office at the Crystal Theater has sold three-fourths of its seats for Saturday's show. The seat map shown has not been updated to show that three-fourths of the available seats have been sold. Draw X's in seats to update the seat map and show exactly three-fourths of the seats sold.

    Sample answer:

    Crystal Theater Seating Map
    ■ Taken
    ☐ Empty

14. Determine whether each statement is true or false.

True	False	
☐	■	$\frac{9}{12}$ is equivalent to $\frac{15}{20}$.
■	☐	$\frac{9}{12}$ is the simplest form of $\frac{15}{20}$.
☐	■	$\frac{9}{12}$ is equivalent to $\frac{18}{24}$.
■	☐	$\frac{3}{4}$ is the simplest form of $\frac{15}{20}$.

15. Roy planted his garden as shown below. Rank Roy's garden from greatest to least with respect to the amount of land each item had.

    - $\frac{2}{5}$ of the garden had tomatoes
    - $\frac{1}{12}$ of the garden had peppers
    - $\frac{1}{6}$ of the garden had lettuce
    - $\frac{2}{10}$ of the garden had herbs
    - $\frac{3}{20}$ of the garden had flowers

    tomatoes, herbs, lettuce, flowers, peppers

Chapter 9 Test

NAME _____ DATE _____ SCORE _____

Chapter 9 Test

1. Write an addition sentence for each model. Then find the sum.

Part A:

$\dfrac{4}{12} + \dfrac{3}{12} = \dfrac{7}{12}$

Part B:

$\dfrac{7}{8} + \dfrac{4}{8} = \dfrac{11}{8} \left(\text{or } 1\dfrac{3}{8}\right)$

Part C:

$1\dfrac{1}{3} + 1\dfrac{2}{3} = 3$

2. Which of the following is equivalent to $\dfrac{1}{2}$? Select all that apply.

■ $\dfrac{5}{8} - \dfrac{1}{8}$ ■ $\dfrac{5}{16} + \dfrac{3}{16}$ ☐ $\dfrac{11}{20} + \dfrac{3}{20}$ ☐ $\dfrac{17}{20} - \dfrac{9}{20}$

3. Sasha and Ali caught fish and weighed them before throwing them back into the water. Sasha's fish weighed $2\dfrac{2}{5}$ pounds. Ali's fish weighed $1\dfrac{4}{5}$ pounds. Draw a model that shows the total weight of both fish.

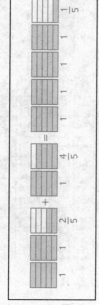

4. Draw a line between each pair of equivalent expressions.

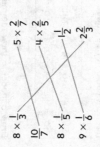

$8 \times \dfrac{1}{3}$ $5 \times \dfrac{2}{7}$

$\dfrac{10}{7}$ $4 \times \dfrac{2}{5}$

$8 \times \dfrac{1}{5}$ $1\dfrac{1}{2}$

$9 \times \dfrac{1}{6}$ $2\dfrac{2}{3}$

ONLINE TESTING On the actual test, you might be asked to click on objects to link them together. In this book, you will be asked to use a pencil to draw a line to show a link.

5. Rosa wrote these addition and subtraction sentences. Describe each mistake that Rosa committed. Correct the mistake. If the sentence is correct, write "Correct."

$\dfrac{3}{5} + \dfrac{2}{5} = \dfrac{5}{10}$ Rosa added the denominators. The correct sum is $\dfrac{5}{5}$ or 1.

$\dfrac{5}{12} - \dfrac{1}{12} = \dfrac{1}{3}$ Correct

$\dfrac{3}{10} + \dfrac{3}{10} = \dfrac{2}{5}$ Rosa simplified incorrectly. The correct sum is $\dfrac{6}{10}$ or $\dfrac{3}{5}$.

$\dfrac{11}{20} - \dfrac{3}{20} = \dfrac{2}{5}$ Correct

Chapter 9 Test

6. The water tank for Camp Utmost is almost full. The head camp counselor needs to file a report with the water department if the tank gets down to half full.

   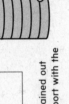

   Part A: What fraction of the tank is now filled? Explain how you found your answer.

   $\frac{9}{10}$ of the tank is filled because the water line is at the 9th of 10 marks.

   Part B: How much more water needs to be drained out of the tank to cause the counselor to file a report with the water department?

   4 more sections, or $\frac{4}{10}$

   Part C: The counselor needs to put the camp on emergency alert if the level gets down to $\frac{1}{5}$ full. From its starting level, how much more water would need to be drained out of the tank to cause an emergency alert to be issued?

   7 more sections, or $\frac{7}{10}$

7. Fill in the missing items to subtract $2\frac{3}{4}$ from $3\frac{1}{4}$. Simplify your answer.

   $3\frac{1}{4} = \frac{4}{4} + \frac{4}{4} + \frac{4}{4} + \frac{1}{4} = \frac{4+4+4+1}{4} = \boxed{\frac{13}{4}}$

   $-2\frac{3}{4} = \frac{4}{4} + \frac{4}{4} + \frac{3}{4} = \frac{4+4+3}{4} = \boxed{\frac{11}{4}}$

   $= \boxed{\frac{2}{4} \text{ or } \frac{1}{2}}$

8. Andrea and Rocky mixed $\frac{4}{6}$ cup of whole wheat flour with $\frac{8}{12}$ cup of rye flour. To find out how much flour they have in all, Rocky said, "We can just add the numerators of the fractions."

   Part A: Is Rocky correct? Explain your answer.

   Rocky is not correct. You can only add like fractions that have the same denominator. These two fractions have different denominators.

   Part B: Andrea had an idea. "Why don't we simplify both fractions. Then we will be able to add them." Is Andrea correct?

   Andrea is correct only if simplifying produces two fractions with the same denominator.

   Part C: Simplify the two fractions. Are you now able to add the two fractions? If so, what sum do you get? If not, explain why the two fractions could not be added.

   Simplifying the fractions gives $\frac{4}{6} = \frac{2}{3}$ and $\frac{8}{12} = \frac{2}{3}$. So $\frac{2}{3} + \frac{2}{3} = \frac{4}{3} = 1\frac{1}{3}$. So Andrea's idea worked.

9. Draw a model to show the multiplication of $3 \times \frac{5}{8}$.

   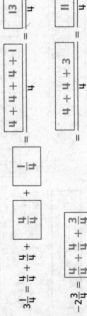

   $\frac{15}{8} = 1\frac{7}{8}$

Chapter 9 Test

10. Select true or false for each statement.

	True	False
$2\frac{4}{9} + 4\frac{8}{9} = 6\frac{1}{3}$	☐	☐
$4\frac{4}{9} - 2\frac{8}{9} = 1\frac{5}{9}$	☐	☐
$6\frac{7}{10} + 6\frac{7}{10} = 13\frac{2}{5}$	☐	☐
$7\frac{7}{10} - 6\frac{9}{10} = 1\frac{8}{10}$	☐	☐

11. Draw a line between each pair of equivalent expressions.

$3\frac{2}{5} + 2\frac{2}{5}$ $5\frac{5}{8} + 3\frac{7}{8}$

4 $\frac{29}{5}$

$3\frac{5}{6} + 3\frac{5}{6}$ $2\frac{5}{12} + 1\frac{7}{12}$

$\frac{45}{8} + \frac{31}{8}$ $7\frac{2}{3}$

12. Joachim started the day with $3\frac{2}{5}$ gallons of paint. He used $1\frac{4}{5}$ gallons to paint the garage door. Then Joachim painted the shed. When he was done painting, Joachim had $\frac{2}{5}$ gallon of paint left. How much paint did Joachim use for the shed? Show how you got your answer.

Total − gallons used for door = Gallons left after door
$3\frac{2}{5} - 1\frac{4}{5} = \frac{17}{5} - \frac{9}{5} = \frac{8}{5}$

Gallons left after door − leftover = Gallons for shed
$\frac{8}{5} - \frac{2}{5} = \frac{6}{5} = 1\frac{1}{5}$

$1\frac{1}{5}$ gallons

13. Use the models to find the sum of the two quantities. Fill in the boxes. Shade the circles and draw new circles if necessary.

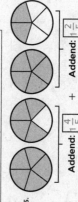
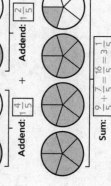

Addend: $1\frac{2}{5}$

Addend: $1\frac{4}{5}$

Sum: $\frac{9}{5} + \frac{7}{5} = \frac{16}{5} = 3\frac{1}{5}$

14. Mona swam back and forth across the Narrow River 8 times. The Narrow River measures $\frac{1}{5}$ of a mile across. Elise swam $\frac{2}{3}$ of the way across the Wide River that measures 5 miles across.

Part A: Which expressions show how far Mona swam? Select all that apply.

☐ $8 \times \frac{1}{5}$
☐ $16 \times \frac{1}{5}$
☐ $\frac{2}{5} \times 8$
☐ $\frac{2}{5} \times 16$

Part B: Which expressions show how far Elise swam? Select all that apply.

☐ $\frac{2}{3} \times 5$
☐ $10 \times \frac{1}{3}$
☐ $20 \times \frac{1}{3}$
☐ $10 \times \frac{2}{3}$

Part C: Who swam farther, Mona or Elise? Explain your answer.

Mona swam $\frac{16}{5}$, or $3\frac{1}{5}$ miles. Elise swam $\frac{10}{3}$, or $3\frac{1}{3}$ miles. $3\frac{1}{3}$ is greater than $3\frac{1}{5}$, so Elise swam farther.

15. The hike from Ducktown to Goose Mountain is $7\frac{4}{5}$ miles. Jim and Becky hiked $1\frac{2}{5}$ miles toward Goose Mountain. Then Jim forgot his water so they hiked back to Ducktown. Then Jim and Becky hiked all of the way to Goose Mountain.

Part A: Write an expression to show far Jim and Becky hiked in all.

Total = $1\frac{2}{5} + 1\frac{2}{5} + 7\frac{4}{5}$

Part B: Jim claims that he and Becky hiked more than 10 miles. Is Jim correct? Explain.

Yes, $1\frac{2}{5} + 1\frac{2}{5} + 7\frac{4}{5}$ is equal to $10\frac{3}{5}$ miles, which is greater than 10.

Chapter 10 Test

NAME _____ DATE _____ SCORE _____

Chapter 10 Test

1. Mindy bought a blue marker for 0.93 dollars. Which coins should she take out to pay for the marker? Draw quarters, dimes, nickels, and pennies in the box. Label each coin.

ONLINE TESTING On the actual test, you may drag coins into the box to show your total. In this book, you will use a pencil to draw in and label the coins that you use.

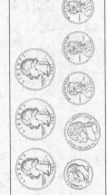

2. Decimal models and a number line are shown.

Part A: Draw a line from each model to its location on the number line.

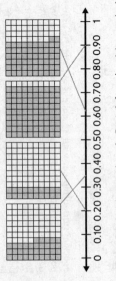

Part B: Write the numbers in **Part A** from least to greatest.

0.2, 0.35, 0.62, 0.9

3. Marques claimed he could write a number or model that is equivalent to 0.70 in six different ways. Is Marques correct? Explain your answer.

Sample answer: Marques is correct. He can show 0.70 as: 0.70, 0.7, $\frac{7}{10}$, $\frac{70}{100}$, a model showing 7 out of 10 shaded bars, or a model showing 70 out of 100 shaded squares. In words, he can write seventy hundredths or seven tenths.

4. Shondra made this model to show the sum of $\frac{6}{10}$ and $\frac{3}{100}$.

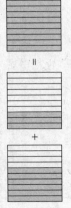

Part A: Is Shondra's model correct? Explain.

Shondra's model is not correct. Her model of $\frac{3}{100}$ is showing $\frac{3}{10}$ instead of $\frac{3}{100}$.

Part B: Draw a correct model for sum of the two fractions.

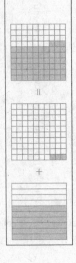

Part C: What is the sum $\frac{6}{10} + \frac{3}{100}$ in decimal form?

0.63

Chapter 10 Test

5. A jeweler melted down a silver nugget and an old necklace to obtain two batches of pure silver. The silver from the nugget weighed 0.7 ounces. The silver from the necklace weighed $\frac{70}{100}$ of an ounce. Which piece had greater value? Explain.

Sample answer: Both pieces have the same value because 0.7 is equal to 0.70, which is equal to $\frac{70}{100}$.

6. Jay's recipe for making tortillas calls for 0.6 pounds of corn meal and 0.19 pounds of vegetable oil.

Part A: Which ingredient is greater in weight? Explain.

The corn meal weighs 0.6 pounds, which is equal to 0.60. 0.60 is greater than 0.19.

Part B: Jay adds both ingredients to a bowl. How much do the ingredients in the bowl weigh? Color in the model to show your answer. Write the total in decimal and fraction form.

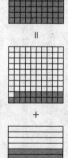

Total: 0.79 or $\frac{79}{100}$ pounds

7. Yasmina is sewing a jacket. For the sleeves Yasmina cuts a piece of cloth that is $\frac{7}{10}$ of a meter in length and 0.2 meter in width. For the front panel Jasmine cuts a piece of cloth that is $\frac{65}{100}$ meter in length and 0.2 meter in width.

Part A: Yasmina compares the lengths of the two pieces. To find out which piece is longer, what information does Yasmina need?

Sample answer: Yasmina needs to know the length of each piece.

Part B: What do you know about the widths of the pieces?

The widths are the same.

Part C: Which piece of cloth is longer? Explain.

Sample answer: The piece for the sleeves is longer because $\frac{7}{10}$ is greater than $\frac{65}{100}$.

8. Belinda made this model of her garden. Each shaded square shows where flowers are planted. Non-shaded squares show the walkways in Belinda's garden.

Part A: What part of the garden has flowers? Write your answer in decimal form.

0.42

Part B: What part of the garden marks represents walkway? Write your answer in decimal form.

0.58

9. Use the model to show a different arrangement of flowers for Belinda's garden in problem 8. The arrangement should have the same amount of space for flowers as Belinda's garden. Explain why your model is equivalent to Belinda's model.

Sample answer: My model is equivalent because it has the $\frac{42}{100}$ of the area shaded, the same area for flowers as Belinda's garden.

Grade 4 • Chapter 10 Fractions and Decimals

Chapter 10 Test

10. At the tech store, Ky saw a case for her phone that cost $4.27. Fill in the table to show the money Ky needs for the phone case.

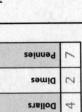

Dollars	Dimes	Pennies
4	2	7

11. Quinn collected $\frac{26}{100}$ of a kilogram of raspberries. She worked from 9:30 A.M to 1:00 P.M. Mitch collected 0.4 kilograms of raspberries. With the information given, which of the following problems can you solve? Select all that apply.

☐ When did Mitch stop collecting raspberries?
☒ How many kilograms did Mitch and Quinn pick together?
☒ How many kilograms did Quinn pick by 12 noon?
☐ Who picked more raspberries, Quinn or Mitch?

12. Arrange these numbers in the box from least to greatest.

0.22 $\frac{2}{10}$ 0.02   0.21 $\frac{23}{100}$ 0.21

| 0.02 | $\frac{2}{10}$ | 0.21 | 0.22 | $\frac{23}{100}$ |

13. Noah ran $\frac{3}{10}$ of a mile to Unger Park. In the park, Noah walked for 0.9 mile with his friend Oswaldo. Finally, Noah and Oswaldo ran 0.48 mile to Oswaldo's house.

Part A: How far did Noah run in all?

0.78 mile

Part B: Did Noah run farther than he walked? Explain.

Sample answer: Noah walked farther than he ran because 0.9 miles is greater than the 0.78 miles that Noah ran.

14. Draw a line between each pair of equivalent expressions.

four tenths — 0.04
$\frac{77}{100}$ — 0.70
four hundredths — 0.40
— 0.77
0.7

15. Select true or false for each statement.

True	False	
☐	☐	$0.1 + \frac{1}{10} = 1.1$
☐	☒	$0.2 + \frac{2}{10} = 0.22$
☐	☐	$0.03 + \frac{3}{10} = 0.33$
☐	☒	$0.04 + \frac{4}{100} = 0.08$

16. Danny says he can show this amount of money using just a single type of coin. Which coin does he need to use and how many of these coins will he need? What is the total number of cents?

Danny needs 13 nickels to make 65 cents.

Chapter 11 Test

NAME _____ DATE _____ SCORE _____

Chapter 11 Test

1. An experimental car that runs on solar power weighs $2\frac{1}{4}$ tons. How many pounds does this car weigh?

   4,500 pounds

2. Compare the horizontal line segments for the two figures.

   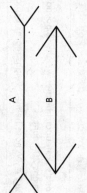

   Part A: Which horizontal line is greater in length? Make an estimate.

   Answers will vary. Most students will say that A is greater in length.

   Part B: Measure each horizontal line segment to the nearest $\frac{1}{4}$ inch. Which figure is greater in length?

   Both figures measure exactly $2\frac{1}{2}$ inches in length.

   Part C: How accurate was your estimate? Explain.

   Sample answer: I thought figure A was greater in length, but it turned out to have the same length as figure B. I think that the "tails" on figure A made it seem longer than it actually was.

3. Elle poured 1 quart, 1 pint, and 1 cup full of water into an empty bucket. How many additional cups of water will it take to fill the 1-gallon bucket?

   9 cups

4. Raymond's recipe for tortilla soup calls for filling this cook pot with broth, black beans, and tomatoes. Raymond does not know how much the pot holds. What would be a good estimate for the capacity of the pot? Select all that apply.

   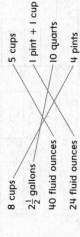

   ☐ 9 gallons
   ☒ $2\frac{1}{2}$ quarts
   ☐ 5 pints
   ☐ 80 fluid ounces

5. Coach Ruiz tells says his football team gained $1\frac{1}{8}$ mile in total yards from scrimmage this season. How many total yards did the team gain?

   1,980 yards

6. Draw a line between each pair of equivalent quantities.

   8 cups — 5 cups
   $2\frac{1}{2}$ gallons — 1 pint + 1 cup
   40 fluid ounces — 10 quarts
   24 fluid ounces — 4 pints

7. Ms. Stephanos bought a house with a 30-year mortgage. She will be making the same payment each month for the next 30 years.

   Part A: How many monthly payments will Ms. Stephanos make? Explain.

   $360 = 30 \times 12$

   Part B: How many weeks will be covered over the 30 year period? Explain.

   $1,560 = 52 \times 30$

Chapter 11 Test

8. Wilson measured the speed of his tennis serve and rounded each speed to the nearest 5 miles per hour.

45 mph	50 mph	55 mph	60 mph	65 mph	70 mph
1	3	5	4	0	1

```
            X
      X   X X
    X X   X X X
  X X X   X X X   X
  +---+---+---+---+---+---+→
  45  50  55  60  65  70
```

Part A: To make a line plot using Wilson's data, you first need to make a scale. Which high and low values and intervals should you choose for the scale? Draw in the values and intervals on the scale.

Sample answer: Low: 45 mph, High: 70 mph

Part B: Use the data to draw in the values for the line graph.

Part C: What was the speed of a typical serve for Wilson? Explain.

Sample answer: A typical serve would be about 55 mph because that value had the greatest number of serves as shown on the graph.

9. Boris's skateboard helmet weighs $1\frac{3}{8}$ pounds. Marisa's helmet weighs $20\frac{1}{4}$ ounces.

Part A: Whose helmet weighs more? Explain how you know.

Sample answer: Boris's helmet weighs $16 + (16 \times \frac{3}{8})$ = 22 ounces. $20\frac{1}{4}$ is less than 22, so Boris's helmet weighs more.

Part B: How much less does the lighter helmet weigh?

$22 - 20\frac{1}{4} = 1\frac{3}{4}$ ounces less

10. Sandy ran 3 laps around a $\frac{1}{4}$-mile track. Which of the following are equal to how far Sandy ran? Select all that apply.

- ☐ 1,320 yards
- ☐ 4,000 feet
- ☒ 3,960 feet
- ☐ 3,960 inches

11. Ephraim estimated the distance that he hit a golf ball. Place a value from the list below inside each box.

| 176 | 6,336 | $\frac{1}{10}$ | 528 |

feet	yards	inches	miles
528	176	6,336	$\frac{1}{10}$

ONLINE TESTING
On the actual test, you may drag items into boxes. In this book, you will write the values in the boxes using a pencil.

12. A blue whale's tongue weighs about $3\frac{1}{4}$ tons. A full-grown hippopotamus weighs 6,150 pounds.

Part A: Which weighs more, the blue whale's tongue or the hippopotamus? Explain how you know.

Sample answer: $3\frac{1}{4}$ tons = $(3 \times 2,000)$ + $(2,000 \times \frac{1}{4})$ = 6,500. 6,500 is greater than 6,150, so the whale's tongue is heavier.

Part B: How much more does the heavier item in **Part A** weigh?

$6,500 - 6,150 = 350$ pounds

Chapter 11 Test

16. Select true or false for each statement.

	True	False	
	☐	☐	7 cups > 2 quarts
	☐	☐	2 quarts > 1 pint + 42 fluid ounces
	☐	☐	$2\frac{1}{8}$ gallon = 17 pints
	☐	☐	$\frac{1}{8}$ quart < $\frac{3}{4}$ cup

17. Nilsa and her classmates collected donations to get new computers for her school. Donations were made in increments of $10. Nilsa made this line graph of the data, showing donations in dollars.

Part A: What was the most common amount that people donated?

$20

Part B: Which donation amount produced the greatest donation total? What was the total donated at that amount?

$50; a total of $150

18. Franklin wants to use a bowling ball that weighs no more than 224 ounces. The label on the box for a new bowling ball shows that the ball weighs 12 pounds 14 ounces. Is this a ball that Franklin could use? Explain.

Yes; 12 pounds 14 ounces = 12 × 16 + 14 = 206 ounces and 206 < 224

---

13. Write the measurements from the list in order from least to greatest.

| $\frac{1}{4}$ quart | $\frac{1}{8}$ gallon | 3 fluid ounces | $1\frac{1}{2}$ pint | $\frac{1}{4}$ cup |

3 fluid ounces, $\frac{1}{2}$ cup, $\frac{1}{4}$ quart, $\frac{1}{8}$ gallon, $1\frac{1}{2}$ pint

14. Davin ran the marathon in $4\frac{3}{5}$ hours. Melba ran the marathon in 266 minutes. After she crossed the finish line, Melba waited for Davin to arrive. How many seconds did she wait? Explain.

600 seconds; Sample answer: $4\frac{3}{5}$ hours = (4 × 60) + ($\frac{3}{5}$ × 60) = 240 + 36 = 276 minutes. 276 − 266 = 10 minutes. 10 × 60 = 600 seconds. Melba crossed the line 10 minutes, or 600 seconds, ahead of Davin.

15. LeVar has 9 coins in his pocket that are worth a total of 86 cents. LeVar has at least one of each of the following kinds of coin: penny, nickel, dime, and quarter. LeVar has more nickels than dimes. What coins does LeVar have? Explain.

Sample answer: I used guess and check.
First I tried 3 quarters (value too high, not enough coins):
(3 × 25) + (1 × 10) + (1 × 5) + (1 × 1) = 91¢, 6 coins
Then 2 quarters (correct value, not enough coins):
(2 × 25) + (2 × 10) + (3 × 5) + (1 × 1) = 86¢, 8 coins
Then 1 quarter (correct value; not enough nickels):
(1 × 25) + (5 × 10) + (2 × 5) + (1 × 1) = 86¢, 9 coins
Then 2 quarters again (correct):
(2 × 25) + (1 × 10) + (5 × 5) + (1 × 1) = 86¢, 9 coins

Grade 4 • Chapter 11 Customary Measurement

Chapter 12 Test

NAME _____ DATE _____

SCORE _____

Chapter 12 Test

1. Select all of the items that would have a volume of about 0.2 L.

   ☐ cup of hot tea
   ☒ jug of ice tea
   ☒ bottle of lotion
   ☐ sip of soup

2. Jerrilyn's wrist watch weighs 0.15 kilograms. Josh's watch weighs 145 grams.

   Part A: Whose watch is heavier? Explain.

   0.15 kilograms = 150 grams. So Jerrilyn's 150 g watch is heavier than Josh's 145 g watch.

   Part B: In grams, how much greater in weight is the heavier watch?

   150 g − 145 g = 5 grams

3. A plan for a fountain is shown. The diameter of the outer circle is 8 meters. The distance between the outer circle and the shaded inner circle measures 2.5 meters. What is the diameter of the shaded inner circle in centimeters? Show your work.

   300 cm;
   8 m − 2.5 m − 2.5 m = 3 m
   3 m = 300 cm

4. In Science lab, Regan needs to fill a tank with 2.25 liters of distilled water. How many times will Regan need to refill a 250 mL beaker with water in order to get the 2.25 liters that she needs?

   9 times

5. Which units would this conversion table fit? Select all that apply.

?	?	?
3	3,000	(3, 3,000)
5	5,000	(5, 5,000)
7	7,000	(7, 7,000)

   ☐ meters and centimeters
   ☒ liters and milliliters
   ☒ kilograms and grams
   ☐ centimeters and kilometers

6. Tracee has 3 songs on her new playlist with titles "Always," "Before I Leave," and "Come Home." On her playlist, Tracee abbreviates the songs as A, B, and C. In how many different orders can Tracee play the 3 songs? Show your work.

   Sample answer: I used Make an Organized List. There are 6 different orders.

   A, B, C   B, A, C   C, A, B
   A, C, B   B, C, A   C, B, A

7. Monty measured the distance across the Juniper Park swimming pool. To show this distance, place a value from the list below inside each box.

cm	km	mm	m
3,200	0.032	32,000	32

   3,200   32   0.032   32,000

   ONLINE TESTING: On the actual test, you may drag items into boxes. In this book, you will write the values in the boxes.

Chapter 12 Test

8. Fresh mozzarella cheese from Antonio's Market costs $4.00 per kilogram.

   Part A: Roxanne paid $8.00 for her mozzarella. Did she buy enough cheese to make one large pizza? Explain.

   Sample answer: Yes, Roxanne bought 2 kilograms. A single pizza would require much less than 2 kilograms of cheese.

   Part B: Roxanne bought 2 large watermelons from Antonio's market. Are the watermelons likely to weigh more or less than the cheese she bought?

   Sample answer: More; Roxanne bought 2 kg of cheese. Two watermelons would likely weigh much more than 2 kg.

9. Draw a line between each pair of equivalent quantities.

   250 mm — 25,000 mm
   250 cm — 0.25 m
   0.25 km — 2.5 m
   25 m — 250 m

10. Select true or false for each statement.

	True	False
250 g > 2.5 kg	☐	☐
3.1 L < 400 mL + 2 L	☐	☐
150 mm > 5 cm	☐	☐
0.4 km < 4,000 m	☐	☐

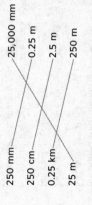

11. In darts, Alison scored 57 points on 8 throws. She hit each score on the board at least once.

    Part A: What scores did Alison get on her 8 throws? Show how you got your answer.

    Sample answer: I used the make an organized list strategy.

    $(2 \times 20) + (1 \times 10) + (1 \times 5) + (2 \times 1) = 57$
    (correct total, wrong number of throws)

    $(1 \times 20) + (3 \times 10) + (1 \times 5) + (2 \times 1) = 57$
    (correct total, wrong number of throws)

    $(1 \times 20) + (2 \times 10) + (3 \times 5) + (2 \times 1) = 57$
    (correct, correct number of throws)

    Part B: Draw the darts on the dartboard.

12. On a metric scale, Ira's fish weighed 1 kilogram. On a customary scale, the same fish weighed a little over 2 pounds.

    Part A: Ira caught a fish that weighed 4,000 grams. About how much would this fish weigh in pounds?

    Sample answer: I made an organized list. The fish would weigh about 8 pounds. 4,000 g = 4 kg
    1 kg = about 2 lb
    2 kg = about 4 lb
    3 kg = about 6 lb
    4 kg = about 8 lb

    Part B: Which would weigh more, a 10 kg fish or a fish that weighed 15 pounds? Explain.

    Sample answer: The 10 kg fish would weigh about 2 pounds per kilogram, or at least 20 pounds, so it would be heavier.

Chapter 12 Test

13. At the county track meet, these were the top 5 finishers in the long jump event. Rank each finisher from first to fifth place.

Name	Distance	Place
Brian	4.08 m	2
Shondra	403 cm	3
Dave	399.9 cm	5
Kayla	4.1 m	1
Hernán	4,000 mm	4

14. Rita is organizing the Hartley Park Treasure Hunt, in which students hunt for buried treasure in the park. Hartley Park is 0.1 kilometers wide and 100 m in length.

Part A: Rita's directions are shown. Circle the most likely choice of unit for each measurement.

1. Start at the main gate of the park. Walk 0.06 (millimeters, centimeters, meters, (kilometers)) north past the playground to the drinking fountain.

2. From the drinking fountain, walk 35 (millimeters, centimeters, (meters), kilometers) through the playground west to the fence.

3. From the fence, walk 2,500 (millimeters, (centimeters), meters, kilometers) south through the hopscotch court to the creek.

4. At the creek, turn and walk about 5 steps, or 3,000 ((millimeters), centimeters, meters, kilometers) north. You will find the treasure there.

Part B: To find the treasure, what total distance do you need to walk? Show your work.

Sample answer: Add the following:
1. 0.06 km = 60 m
2. 35 m
3. 2,500 cm = 25 m
4. 3,000 mm = 3 m
Total = 60 + 35 + 25 + 3 = 123 m

15. The figure is placed next to a centimeter ruler.

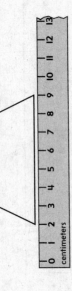

Part A: What is the approximate length of the bottom side of the figure? Select all that apply.
- [ ] 9 cm
- [ ] 70 mm
- [ ] 700 mm
- [ ] 0.07 m

Part B: What is the approximate length of the top side of the figure? Select all that apply.
- [ ] 50 mm
- [ ] 0.05 m
- [ ] 8 cm
- [ ] 0.5 cm

16. At the beach, Edison sells sunglasses and tubes of sunblock. He orders 15 pairs of sunglasses that weigh 26 grams each. He also orders 12 tubes of sunblock that weigh 43 grams each.

Part A: Is the total weight of Edison's order greater than or less than 1 kilogram? Explain.

Sample answer: sunglasses: 15 × 26 = 390,
sunblock 12 × 43 = 516
390 + 516 = 906 g, which is less than 1,000 g, or 1 kg

Part B: How many more pairs of sunglasses would Edison need to add or subtract in order to have his total order weigh about 1 kilogram?

about 4 more pairs of sunglasses

Chapter 13 Test

NAME _____ DATE _____ SCORE _____

Chapter 13 Test

1. The rectangle has a length of 8 centimeters and an area of 24 square centimeters.

   8 cm / ?
   Area = 24 sq cm

   Part A: What is the width of the rectangle?

   3 cm

   Part B: Draw in the square unit centimeters to show the area of the figure.

2. What is the next figure in this pattern? Draw the figure. What is the area of the figure?

   30 square units

3. The length of a rectangle is twice the width of the rectangle. The perimeter of the rectangle is 36 inches. Complete the table.

Length (in.)	Width (in.)	Perimeter (in.)	Area (sq. in.)
12	6	36	72

   ONLINE TESTING
   On the actual test, you may drag items into boxes. In this book, you will write the values in the boxes.

4. Draw a line between each pair of matching items.

   square feet — length × width
   area of rectangle — perimeter units
   inches — (length × 2) + (width × 2)
   perimeter of rectangle — area units

5. Vanessa is an architect. She designed Goldfish Pond I for her client as a square.

   Goldfish Pond 1: 16 ft × 16 ft
   Goldfish Pond 2: with 4 ft square cut-out, 6 ft, 6 ft, 4 ft, 16 ft

   Part A: Find the perimeter of Goldfish Pond I.

   64 ft

   Part B: Vanessa changed Goldfish Pond I into Goldfish Pond 2. Pond 2 has a square cut-out that measures 4 feet on each side. What is the perimeter of Pond 2? Show your work.

   P = 16 + 16 + 16 + 6 + 4 + 4 + 4 + 6 = 72 ft

   Part C: Compare the perimeters of Pond 1 and Pond 2. Why are they different? Explain.

   Pond 2 has a greater perimeter. The cut-out increased the perimeter.

6. Vanessa from Problem 5 made a sketch of Goldfish Pond H.

   Goldfish Pond H, 16 ft

   Part A: Without knowing the size of the cut-outs, what can you say about the perimeter of Pond H compared to Pond I and Pond 2? Which pond will have the greatest perimeter? Which will have the least perimeter? Explain.

   Pond H will have a greater perimeter than Pond I or Pond 2. Pond I will have the least perimeter. Creating cut-outs increases the perimeter.

   Part B: Change Pond H so that its perimeter increases. Make a drawing.

   Sample answer:
   16 ft

Chapter 13 Test

10. Jazmine created a floor plan for her "dream" kitchen.

    Jazmine's Dream Kitchen
    (Pantry 6 ft × 10 ft; Cooking area 14 ft × 14 ft with 2 ft notch; Nook 14 ft × 9 ft with 8 ft top)

    Part A: In order to find the area of her dream kitchen, Jazmine divided the kitchen into three areas: the Cooking Area, the Pantry, and the Nook. What is the total area of Jazmine's dream kitchen?

    $A = (10 \times 6) + (14 \times 8) + (14 \times 9) = 298$ square feet

    Part B: What is the perimeter of Jazmine's dream kitchen?

    $P = 29 + 10 + 6 + 2 + 14 + 6 + 9 + 14 = 90$ ft

    Part C: To check her perimeter answer, Jazmine found the sum of the perimeters of the three areas. Does the sum of the perimeters of the three areas agree with the perimeter you found in **Part B**? If so, explain why. If not, explain which value for the perimeter is correct.

    Sample answer: P (3 areas) $= 2 \times (10 + 6) + 2 \times (14 + 8) + 2 \times (14 + 9) = 122$ ft. No, the sum of the perimeters of the individual areas is 122 ft, much greater than the kitchen perimeter of 90 ft. The sum of perimeters for the 3 areas is greater because when you separate the areas you counted "inside" walls that are not part of the kitchen perimeter.

---

7. Carlos bought 28 meters of fencing to create a dog run for his dog. The dog run must be measured in whole meters.

    Part A: Fill in the table. Each dog run arrangement must use all of the fencing that Carlos bought.

Length (ft)	Width (ft)	Perimeter (ft)	Area (sq ft)
12	2	28	24
11	3	28	33
10	4	28	40
9	5	28	45
8	6	28	48
7	7	28	49

    Part B: To give his dog the greatest area for play, which design should Carlos choose? Explain.

    Carlos should use the 7 m by 7 m square because it provides the greatest area of 49 sq ft and still has a 28 m perimeter.

8. Carlos in Problem 7 created a dog run for Mr. Escobar using 40 feet of fencing. Mr. Escobar's dog run has 100 square feet of space. What are the dimensions of Mr. Escobar's dog run? Explain.

    10 ft by 10 ft

9. The figure is made of small squares that measure 3 cm on a side. What is the area of the shaded part of the figure?

    Each square $= 3 \times 3 = 9$ sq cm
    24 shaded squares: $24 \times 9 = 216$ sq cm

266 Grade 4 • Chapter 13 Perimeter and Area

Chapter 13 Test

11. Paul is putting in a 36 square foot model train course in his basement. The course must have a rectangular shape and be measured in whole meters. The course can have any length but it must be at least 2 meters in width.

Part A: What are the possible dimensions for the course?

2 × 18, 3 × 12, 4 × 9, 6 × 6

Part B: Paul wants the trains to have the longest course possible and still take up only 36 square feet of space. Which dimensions for the course should he choose? Explain.

Sample answer: 2 × 18; This arrangement has the greatest perimeter of any of the choices, 40 feet.

12. The swimming pool is surrounded by a 1-meter walkway. Compare the area of the walkway and the area of the pool. Which area is greater? By how much is it greater?

Pool area: 18 sq. m, walkway area: 22 sq m; so the walkway area is greater by 4 sq m

13. Select true or false for each statement about P, the perimeter of a figure.

True False
☐   ■ $P = 4 \times$ side for all squares
■   ☐ $P = 4 \times$ side for all rectangles
■   ☐ $P = (2 \times$ length$) + (2 \times$ width$)$ for all squares
☐   ■ $P = (2 \times$ length$) + (2 \times$ width$)$ for all rectangles

14. A rectangular indoor running track measures 60 yards by 52 yards. Giorgi ran 1,500 yards around the track. How many laps did he finish? Make a table to solve the problem.

Lap	Yards
1	224
2	448
3	672
4	896
5	1,120
6	1,344
7	1,568

Giorgi ran 6 full laps and part of another lap.

15. Rectangle A measures 6 cm by 4 cm. Rectangle B has the same area as Rectangle A and is 1 cm in width. Which of the following is true? Select all that apply.

☐ The perimeter of Rectangle A is greater.
■ The perimeter of Rectangle B is greater.
☐ Rectangle B is greater in length than Rectangle A.
☐ Rectangle A is greater in length than Rectangle B.

16. Match the dimensions of each rectangle with the correct area and perimeter.

4 ft × 6 ft $P = 20$ ft, $A = 16$ sq ft
5 ft × 5 ft $P = 20$ ft, $A = 24$ sq ft
8 ft × 2 ft $P = 28$ ft, $A = 24$ sq ft
12 ft × 2 ft $P = 20$ ft, $A = 25$ sq ft

NAME _____ DATE _____

SCORE _____

Chapter 14 Test

1. Consider the figure shown.

   Part A: Which line segments are parallel? Select all that apply.
   - ☐ $\overline{AG}$ and $\overline{AB}$
   - ☐ $\overline{FE}$ and $\overline{CD}$
   - ☐ $\overline{FB}$ and $\overline{FC}$
   - ☐ $\overline{AG}$ and $\overline{FE}$

   Part B: Which line segments are perpendicular? Select all that apply.
   - ☐ $\overline{AG}$ and $\overline{AB}$
   - ☐ $\overline{GC}$ and $\overline{CD}$
   - ☐ $\overline{FB}$ and $\overline{GF}$
   - ☐ $\overline{GF}$ and $\overline{FE}$

2. Draw a line between each pair of matching items.

   obtuse angle — right angle
   acute angle — $\frac{1}{6}$ turn
   $\frac{1}{4}$ turn — 180°
   straight angle — greater than 90°

3. Draw an obtuse triangle with two equal acute angles.

4. A billiard ball travels in the direction shown, bounces off a wall at angle a and rebounds at the exact same angle. The angle between the incoming path and outgoing path of the ball is 120°.

   Part A: What is the sum of the measures of the three angles shown?

   180°

   Part B: What is the measure of angle a?

   30°

5. Draw lines of symmetry for each letter. Circle the letters that do not have a line of symmetry.

   H E L P N O W

6. Rex claims he can draw a right triangle that has one right angle, one acute angle, and one obtuse angle. Is Rex correct? Explain.

   Sample answer: Rex is not correct. A right triangle must have one right angle and two acute angles. A right triangle cannot have both a right angle and an obtuse angle.

268 Grade 4 · Chapter 14 Geometry

Chapter 14 Test

7. Three angles combine to form a straight angle. Which of the following must be true? Select all that apply.
   - [ ] One of the angles must be obtuse.
   - [x] At least two of the three angles must be acute.
   - [ ] All three angles must be acute.
   - [ ] None of the angles can be obtuse.
   - [x] All three angles can be acute.
   - [x] One of the angles can be a right angle.
   - [ ] One of the angles must be a right angle.

8. Angles x and y combine to form a 120° angle. The angle measure of y is twice the angle measure of x. What are the measures of angle x and angle y? Show your work.

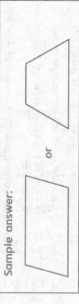

Sample answer: I used guess and check.
If $x = 60°$, then $y = 2 \times 60° = 120°$, so $x + y = 180°$. (too large)
If $x = 35°$, then $y = 2 \times 35° = 70°$, so $x + y = 105°$. (too small)
If $x = 45°$, then $y = 2 \times 45° = 90°$, so $x + y = 135°$. (too large)
If $x = 40°$, then $y = 2 \times 40° = 80°$, so $x + y = 120°$. (correct)
So $x = 40°$, and $y = 80°$.

9. A quadrilateral has two obtuse angles that are equal in measure and two acute angles that are equal in measure. Draw the quadrilateral.

Sample answer:

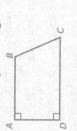 or

10. A quadrilateral has two right angles and one pair of parallel line segments.

    Part A: Draw the quadrilateral and label its right angles. Is the quadrilateral a rectangle or a parallelogram? Explain.

    Sample answer: No, the quadrilateral is neither a rectangle nor a parallelogram. It has only two right angles, not four right angles like a rectangle. It has only one pair of parallel segments, not two pairs like a parallelogram.

    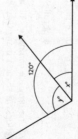

    Part B: Which line segments in the quadrilateral are parallel?

    $\overline{AB}$ and $\overline{DC}$

11. Find the measure of each angle in the quadrilateral. What is the sum of all four angles?

    angle A: 40°, angle B: 133°, angle C: 122°, angle D: 65°; Sum of angles: 360°

Chapter 14 Test

12. Warren drew a diagonal line to connect two opposite corners of a rectangle. Then he cut the rectangle in two along that diagonal line. What figures were formed? Which figure is greater in size? Use a diagram to explain your answer.

Two right triangles are formed that are equal in size.

13. Write the number of lines of symmetry that each figure has in the box.

an oval	2
a person	1
a glove	0
the letter P	0
an 8-sided stop sign	8
a parallelogram	0

14. Will the gray tile completely fill in the shaded space without any gaps or overlaps? If so, how many tiles will it take? Make a model to obtain your answer.

10 tiles

15. Which quadrilaterals can have at least one pair of parallel sides, at least one pair of sides that are equal in length, and at least one right angle?

rectangle, square, rhombus, trapezoid

16. Dania started with her angle arrow in the position shown. Then she took these four steps:

Step 1: $\frac{1}{4}$ turn left
Step 2: $\frac{1}{2}$ turn right
Step 3: $\frac{1}{4}$ turn right
Step 4: more than $\frac{1}{2}$ turn left

In what position does Dania's angle arrow end up? Draw the arrow.

Sample answer:

17. Write "parallel," "perpendicular," or "neither" for each pair of streets.

perpendicular	Roy Street and Lake Lane
neither	Roy Street and 1st Street
parallel	Pine Street and Roy Street
perpendicular	K Road and 3rd Street
parallel	1st Street and 3rd Street
neither	Lake Lane and 2nd Street

18. Indicate whether each statement is true or false.

True	False	
☐	☑	A half-turn is equal to 180°.
☐	☑	A right triangle has two right angles.
☑	☐	A parallelogram can have a right angle.
☐	☑	A rhombus can be a square, but a square cannot be a rhombus.
☑	☐	A trapezoid always has one pair of sides with equal lengths.

Chapter 1 Performance Task Rubric

Page 137 • Get to Know A National Park

Task Scenario

Students will apply concepts of place value to read, write, compare, and round multi-digit whole numbers that represent data from a national park's statistics.

Depth of Knowledge	DOK2, DOK3

Part	Maximum Points	Scoring Rubric
A	4	Full Credit: A bar graph titled "Number of Visitors" by Month, with bars showing: January ≈ 27,000; February ≈ 28,000; March ≈ 19,000; April ≈ 31,000. Partial Credit (3 points) will be given if January or March are slightly off but still between the respective grid lines. Partial Credit (2 points) will be given for one or two completely incorrect or missing bars. Partial Credit (1 point) will be given for three incorrect or missing bars. No credit will be given if all bars are incorrect or missing.

Chapter 1 Performance Task Rubric, continued

Part	Maximum Points	Scoring Rubric
B	3	**Full Credit:** 312,026 The number is between 300,000 and 400,000, so the digit in the hundred thousands place is 3. The digit 1 is in the thousands place in the April number. In May, it is worth 10 times that and is, therefore, in the ten thousands place. The tens and the thousands digit have to be 2, 4, 6, or 8, but the only one that can be multiplied by 3 to get a single digit for the ones place is 2. So, the tens and the thousands digits are 2, and the ones digit is 6. The remaining digit is in the hundreds place, so that digit is 0. Partial Credit (2 points) will be given for one incorrect digit or for two if the two incorrect ones are in the thousands and tens places and/or for one mistake in the explanation. Partial Credit (1 point) will be given for two incorrect digits or for three if two of the incorrect ones are in the thousands and tens places and/or for two mistakes in the explanation. No partial credit will be given for three or more mistakes in the digits or the explanation.
C	2	**Full Credit:** $3 \times 1,000 + 4 \times 100 + 6 \times 10 + 8 \times 1$ three thousand, four hundred sixty-eight Partial Credit (1 point) will be given for having either of the two forms written correctly.
D	3	**Full Credit:** The Needle (9,862) < Grizzly Peak (9,915) < Mount Chittenden (10,088) < Hoyt Peak (10,344) < Saddle Mountain (10,394) < Castor Peak (10,804) (Students may list the mountain name or the height, but do not have to list both.) Student provides a correct explanation of how to compare the mountain heights, referring to place value and comparing digits. Partial Credit (2 points) will be given for one error/omission in ordering and/or in the explanation, including if all of the symbols are in the wrong order but there are no other mistakes. Partial Credit (2 points) will be given for two errors/omissions in ordering and/or in the explanation **OR** if the student orders the mountains correctly from greatest to least with no other errors. No credit will be given for three or more errors/omissions in ordering and/or in the explanation.
TOTAL	12	

Chapter 1 Performance Task Student Model 1

NAME _____ DATE _____

SCORE _____

Performance Task

Get to Know a National Park

Anna is doing a school project on Yellowstone National Park. She also plans to visit the park with her family this summer.

Write your answers on another piece of paper. Show all your work to receive full credit.

Part A

The table shows the number of visitors that Yellowstone has had during the first five months of the year. Draw a bar graph below the table that shows the number of visitors for each month rounded to the nearest thousand.

Month	Number of Visitors to Yellowstone
January	26,778
February	28,233
March	18,778
April	31,356

[Bar graph drawn by student showing Number of Visitors (16,000 to 34,000) for January, February, March, April]

Performance Task (continued)

Part B

In May, between 300,000 and 400,000 people visited Yellowstone. In the total number of visitors for May, the digit 1 has 10 times the value that it has in the total number of visitors in April. In the total number of visitors for May, the tens digit and the thousands digit are both the same even number, and the ones digit is three times the tens digit. The only other digit is a zero, and it is the only zero in the number. How many people visited Yellowstone in May? Explain your reasoning.

(student notes: 3 hundred thousands, 10,000, ×10 ×3)

312,026

Part C

Yellowstone covers 3,468 square miles of land. Write the expanded form and the word form for Yellowstone's land area.

3×1,000 + 4×100 + 6×10 + 8×1

three thousand, four hundred sixty-eight

Part D

The table shows the heights of several mountains in the park.

Mountain	Height (feet)	
The Needle	9,862	1
Hoyt Peak	10,344	4
Saddle Mountain	10,394	5
Grizzly Peak	9,915	2
Mount Chittenden	10,088	3
Castor Peak	10,804	6

Order the mountains from least to greatest according to height. Use inequality symbols to compare the mountains by height. Explain how you used place value to order the mountains.

Needle < Grizzly < Chittenden < Hoyt < Saddle < Castor

I compared place value from left to right.

Chapter 1 Performance Task Student Model 2

NAME _____ DATE _____ SCORE _____

Performance Task

Get to Know a National Park

Anna is doing a school project on Yellowstone National Park. She also plans to visit the park with her family this summer.

Write your answers on another piece of paper. Show all your work to receive full credit.

Part A

The table shows the number of visitors that Yellowstone has had during the first five months of the year. Draw a bar graph below the table that shows the number of visitors for each month rounded to the nearest thousand.

Month	Number of Visitors to Yellowstone
January	26,778
February	28,233
March	18,778
April	31,356

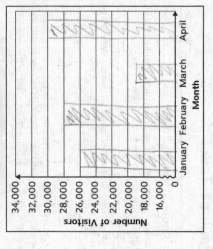

Performance Task (continued)

Part B

In May, between 300,000 and 400,000 people visited Yellowstone. In the total number of visitors for May, the digit 1 has 10 times the value that it has in the total number of visitors in April. In the total number of visitors for May, the tens digit and the thousands digit are both the same even number, and the ones digit is three times the tens digit. The only other digit is a zero, and it is the only zero in the number. How many people visited Yellowstone in May? Explain your reasoning.

312,026

Part C

Yellowstone covers 3,468 square miles of land. Write the expanded form and the word form for Yellowstone's land area.

3×1,000 + 4×100 + 6×10 + 8×1

Part D

The table shows the heights of several mountains in the park.

Mountain	Height (feet)
The Needle	9,862
Hoyt Peak	10,344
Saddle Mountain	10,394
Grizzly Peak	9,915
Mount Chittenden	10,088
Castor Peak	10,804

Order the mountains from least to greatest according to height. Use inequality symbols to compare the mountains by height. Explain how you used place value to order the mountains.

Compare the ones less than 10,000, then Compare the ones more than 10,000. Check each place value.

Chapter 1 Performance Task Student Model 3

NAME _____ DATE _____ SCORE _____

Performance Task

Get to Know a National Park

Anna is doing a school project on Yellowstone National Park. She also plans to visit the park with her family this summer.

Write your answers on another piece of paper. Show all your work to receive full credit.

Part A

The table shows the number of visitors that Yellowstone has had during the first five months of the year. Draw a bar graph below the table that shows the number of visitors for each month rounded to the nearest thousand.

Month	Number of Visitors to Yellowstone
January	26,778
February	28,233
March	18,778
April	31,356

(bar graph showing Number of Visitors vs. Month: January, February, March, April)

Performance Task (continued)

Part B

In May, between 300,000 and 400,000 people visited Yellowstone. In the total number of visitors for May, the digit 1 has 10 times the value that it has in the total number of visitors in April. In the total number of visitors for May, the tens digit and the thousands digit are both the same even number, and the ones digit is three times the tens digit. The only other digit is a zero, and it is the only zero in the number. How many people visited Yellowstone in May? Explain your reasoning.

313,039

Part C

Yellowstone covers 3,468 square miles of land. Write the expanded form and the word form for Yellowstone's land area.

three thousand, four hundred sixty-eight

Part D

The table shows the heights of several mountains in the park.

Mountain	Height (feet)
The Needle	9,862
Hoyt Peak	10,344
Saddle Mountain	10,394
Grizzly Peak	9,915
Mount Chittenden	10,088
Castor Peak	10,804

Order the mountains from **least to greatest** according to height. Use inequality symbols to compare the mountains by height. Explain how you used place value to order the mountains.

Needle > Grizzly > Chittenden > Hoyt > Saddle > Castor

Chapter 1 Performance Task Student Model 4

NAME _____ DATE _____

SCORE _____

Performance Task

Get to Know a National Park

Anna is doing a school project on Yellowstone National Park. She also plans to visit the park with her family this summer.

Write your answers on another piece of paper. Show all your work to receive full credit.

Part A

The table shows the number of visitors that Yellowstone has had during the first five months of the year. Draw a bar graph below the table that shows the number of visitors for each month rounded to the nearest thousand.

Month	Number of Visitors to Yellowstone
January	26,778
February	28,233
March	18,778
April	31,356

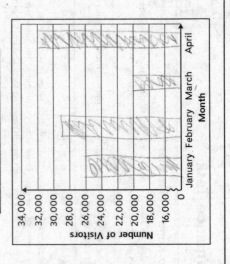

Performance Task (continued)

Part B

In May, between 300,000 and 400,000 people visited Yellowstone. In the total number of visitors for May, the digit 1 has 10 times the value that it has in the total number of visitors in April. In the total number of visitors for May, the tens digit and the thousands digit are both the same even number, and the ones digit is three times the tens digit. The only other digit is a zero, and it is the only zero in the number. How many people visited Yellowstone in May? Explain your reasoning.

314,042

Part C

Yellowstone covers 3,468 square miles of land. Write the expanded form and the word form for Yellowstone's land area.

3×1,000+4×100+6×10+8×1

Part D

The table shows the heights of several mountains in the park.

Mountain	Height (feet)
The Needle	9,862
Hoyt Peak	10,344
Saddle Mountain	10,394
Grizzly Peak	9,915
Mount Chittenden	10,088
Castor Peak	10,804

Order the mountains from least to greatest according to height. Use inequality symbols to compare the mountains by height. Explain how you used place value to order the mountains.

Chapter 2   Performance Task Rubric

Page 139 • The Science Club

Task Scenario

Students will use addition and subtraction to solve multi-step word and analytical problems centered around science club activities; students will compare payment options, plan trip numbers based on restrictions, recognize a pattern, find missing digits in a subtraction problem, and write and solve an equation involving a variable.

Depth of Knowledge	DOK2, DOK3

Part	Maximum Points	Scoring Rubric
A	3	**Full Credit:** Individual Payment: $9 \times 6 + 10 + 10 \times 5 = 54 + 10 + 50 = \$114$ Group Plan: $90 Difference: $114 − $90 = $24 The group plan is better because the trip would cost $114 if the tickets and lunches are purchased individually. That is $24 more than the $90 group plan. Partial Credit (2 points) will be given for one incorrect calculation **OR** if the student calculates correctly but chooses the individual option. Partial Credit (1 point) will be given for two incorrect calculations **OR** if there is one incorrect calculation **AND** the student chooses the individual option. No credit will be given if more than two calculations and the answer to which option to choose are incorrect.
B	1	**Full Credit:** $260 − 33 − 119 − 29 − 38 = 41$ At most 41 students can still sign up to ride the bus. No credit will be given for an incorrect answer.
C	3	**Full Credit:** There are not enough ones to subtract 5 ones from 2 ones, so rename one ten, leaving 7 tens, as 10 ones. 12 ones − 5 ones = 7 ones. 7 tens − x = 4 tens, $x = 4$. 6 hundreds can't be subtracted from 0 hundreds, so rename one thousand, leaving 1 thousand, as 10 hundreds. 10 hundreds − 6 hundreds = 4 hundreds. Partial Credit (2 points) will be given if one digit is incorrect **OR** if the explanation is incomplete or incorrect; the explanation is not complete/correct if it does not refer to place value.

Chapter 2 Performance Task Rubric, continued

Part	Maximum Points	Scoring Rubric
		Partial Credit (1 point) will be given if both digits are incorrect **OR** if one digit is incorrect **AND** the explanation is incomplete or incorrect; the explanation is incomplete/incorrect if it does not refer to place value. No credit will be given for an incorrect answer.
D	2	Full Credit: The difference between the total of Week 2 and Week 1 is $326 − $237 = $89. The difference between the total of Week 3 and Week 2: $415 − $326 = $89. So the amount the science foundation added each week is $89. To find the starting amount (Week 0): $237 − $89 = $148. To find Week 4: $415 + $89 = $504. <table><tr><th>End of Week</th><th>Amount in Fund ($)</th></tr><tr><td>0</td><td>148</td></tr><tr><td>1</td><td>237</td></tr><tr><td>2</td><td>326</td></tr><tr><td>3</td><td>415</td></tr><tr><td>4</td><td>504</td></tr></table> Partial Credit (1 point) will be given if the student makes an error in subtraction but uses that value correctly when calculating the amounts for Week 0 and Week 4. No credit will be given if the student does not subtract to determine the donation amount **OR** if the student uses two different numbers to arrive at the Week 0 and Week 4 amounts.
E	2	Full Credit: The club needs: $3,358 + $1,478 + $600 = $5,436. It already has: $537 in the account + $2,000 from a donor = $2,537. ($3,358 + $1,478 + $600) − ($537 + $2,000) = n (or equivalent) $n = $2,899. The club needs to raise $2,899 more to pay for the trip. Partial Credit (1 point) will be given if the student makes an error in addition or subtraction but sets up the equation correctly. No credit will be given for an incorrect equation.
TOTAL	11	

Part A

If tickets and lunch are bought separately, then there are 8 student tickets that cost $6 each, 1 adult ticket for $10, and 9 lunches for $5 each. 8×6 + 10 + 9×5 = 48 + 10 + 45 = 48 + 55 = 103. The group package costs $96. The group package costs 103 − 90 = $13 less.

Part B

260 total students can ride the bus. Find the number of people signed up and subtract from 260.
33 + 119 = 152
29 + 38 = 67
152 + 67 = 219
260 − 219 = 41

At most 41 students can still sign up to ride the bus.

Part C

I don't have enough ones to take 5 from, so use 1 ten as 10 ones. There are 12 ones and 7 tens. 7 tens minus an unknown number of 10 gives 4 tens, so the unknown tens digit is 3.

Use 1 thousand as 10 hundreds. This leaves 1 thousand and 10 hundreds. 10 − 6 = 4. So the missing digit in the hundreds place is 4.

Part D

326 − 237 = 89   415 − 326 = 89
237 − 89 = 148 (week 0)
415 + 89 = 504 (week 4)

Part E

The club needs 3358 + 1478 + 600 − 537 − 2000 = n for the trip.
5436 − 2537 = n
2899 = n

The club needs to raise $2899 more to pay for the trip.

Chapter 2 Performance Task Student Model 2

Part A
Separate - 8×6 +10 + 9×5 = $103
group - 90
103 - 90 = 13
The group package is less expensive by $13, but the group package is for 10 people, and only 9 are going on the trip, so they should buy everything separately.

Part B
33 + 119 + 29 + 38 = 219
260 - 219 = 41
41 more students can ride the bus.

Part C
I can't take away 5 from 2, so I can regroup to get 12 and cross out the 8 to get 7. 7 - 3 = 4. So the first missing number is 3.
6 can't be subtracted from 0, so I can regroup to get 10. 10 - 6 = 4. So the second missing number is 4.

Part D
Week 2 - Week 1 = 326 - 237 = 89
Week 3 - Week 2 = 415 - 326 = 89
Week 0 = Week 1 - 89 = 237 - 89 = 148
Week 4 = Week 3 + 89 = 415 + 89 = 504

Part E
The club needs 3358 + 1478 + 600 = 5436.
The club has 537 + 2000 = 2537.
5436 - 2537 = n
2899 = n
The club needs $2899 more.

Chapter 2 Performance Task Student Model 3

A) Separate costs: 8×11 +10 = 88+10 = 98.
Group Package: 90
98−90 = 8
The group package is $8 less, but it is for 10 people, and only 9 people are going. They should buy everything separately.

B) The number of people signed up is
33 + 119 + 29 + 38 = 219.
There are 5 buses that seat 260 students each.
260 + 260 + 260 + 260 − 219 = 1081
1081 more students can ride the bus.

C) I can't subtract 5 from 2, so I can regroup to get 12 and cross out 8 to get 7. 7−3=4, so the first missing digit is 3. 6 can't be subtracted from 0, so I can regroup to get 10. 10−6=4, so the second missing digit is 4.

D) 326 − 237 = 111
So Week 0 is: 237 − 111 = 126
Week 4 is: 415 + 111 = 526

E) The club needs 3358 + 1478 + 600 = 5436
They have 537 + 2000 = 2537
5436 − 2537 = (2899)

A. The club should buy everything separately. $8 \times 11 + 10 = 88$. The group package is 90. The difference is $90 - 88 = 2.

B. The number of people signed up is $33 + 119 = 152 + 29 = 181 + 38 = 219$. Subtract $260 - 219 = 41$. 41 more can ride the bus.

C. 8 plus some number ends in 4, so that has to be $8 + 6 = 14$. So the first missing digit is 6. Then $0 + 6$ is also 6, which is the second missing digit.

D. $326 - 237 = 99$
$415 - 326$
Week 0 = $237 - 99 = 138$
Week 4 = $415 + 99 = 414$

E. $3358 + 1478 + 600 - 537 + 2000 = n$
The club needs to raise $6899 more to pay for the trip.

Chapter 3 Performance Task Rubric

Page 141 • Crafty Artists

Task Scenario
Students will use arrays, equations, verbal statements, and number sentences to illustrate properties of multiplication, compare values, demonstrate the relationship between multiplication and division, and illustrate different factor pairs involving craft supply quantities and costs.

Depth of Knowledge	DOK2, DOK3

Part	Maximum Points	Scoring Rubric
A	3	Full Credit: $2 \times 8 \times 10 = 8 \times 10 \times 2$ This demonstrates the Commutative Property of Multiplication because each student does the calculation in a different order but gets the same answer. **OR** $(2 \times 8) \times 10 = 2 \times (8 \times 10)$ This demonstrates the Associative Property of Multiplication because each student groups the factors in a different way. Partial Credit (2 points) will be given for the correct equation and property but no explanation of the property **OR** for the correct equation and explanation of the property but no property name. Partial Credit (1 point) will be given for a correct equation **OR** a correct property explanation but no property name. No credit will be given for only one side of the equation **AND** no property being identified.
B	2	Full Credit: A box of paint markers costs four times as much as a plain t-shirt. $12 = 4 \times 3$ (or equivalent, involving multiplication) Partial Credit (1 point) will be given for a correct comparison statement involving "as much as" or "times" **OR** a correct equation involving multiplication. No credit will be given for no comparison statement **OR** a comparison statement that is incorrect or involves addition **AND** an incorrect equation or one involving no multiplication.

Part	Maximum Points	Scoring Rubric
C	3	Full Credit: $40 \div p = 4$; $p \times 4 = 40$; $p = 10$ (one equation must show multiplication, and the other, division) Partial Credit (2 points) will be given for two correct equations and no solution **OR** for one correct equation and a correct solution. Partial Credit (1 point) will be given for only one correct equation or only the correct solution. No credit will be given for no correct equations **AND** no solution or an incorrect solution.
D	2	Full Credit: 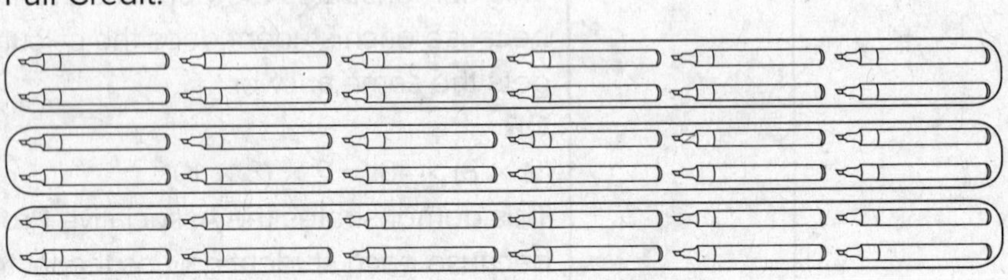 The club ordered 12 boxes of markers in September. Partial Credit (1 point) will be given for the correct circling/grouping **OR** for the correct number of boxes for September. No credit will be given for no or incorrect circling/grouping **AND** for an incorrect number of boxes for September.
E	2	Full Credit: Two different arrays of 24 circles:  Partial Credit (1 point) will be given for only one correct array of 24 circles. No credit will be given for arrays with different numbers of circles in different rows **OR** for arrays with greater or fewer than 24 circles.
TOTAL	12	

Chapter 3 Performance Task Student Model 1

NAME _____ DATE _____ SCORE _____

Performance Task

Crafty Artists

Students in the Future Artists club decorate and sell t-shirts and mugs. The table shows the supplies that the club ordered during November.

Item	Number Ordered	Cost per Item
plain t-shirt	40	$3
plain mug	24	$2
box of fabric markers	24	$10
box of paint markers	36	$12

Write your answers on another piece of paper. Show all your work to receive full credit.

Part A

The club orders an *additional* 2 cartons of fabric markers. Each carton contains 8 boxes of markers. To find the total cost, Ian first calculated 2×8. He then multiplied the result by the cost of each box. Mila also calculated the total. She first found the cost of 8 boxes of markers. She then multiplied the result by the number of cartons. Write a single number statement that shows that Ian and Mila's totals are equal. Explain the property that the number statement demonstrates.

$2 \times 8 \times 6 = 8 \times 10 \times 2$ Comutative property-
multiplying the same numbers in a different order
gives the same answer.

Part B

Write a word statement that compares the cost of a box of paint markers to the cost of a plain t-shirt. Then write a number statement involving multiplication that shows the same comparison.

Paint markers cost 4 times as much as
a t-shirt.
$12 = 4 \times 3$

Performance Task (continued)

$40 \div p = 4$
$p \times 4 = 40$
$p = 10$

Part C

Plain t-shirts come in packages of 4. In order to find p, the number of packages of t-shirts in the order, Jada writes the equation $40 \div 4 = p$. Write two *different* equations, one using multiplication, and the other using division, that could also be used to find p. Then find the value of p.

Part D

The club ordered three times as many boxes of paint markers in November as it did in September. Use the diagram below to show how you can find the number of boxes of paint markers the club ordered in September. Then state the number of boxes ordered in September.

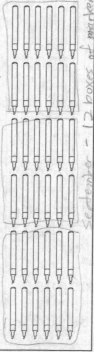

September = 12 boxes of markers

Part E

The club members decorated all of the mugs ordered in November. They arranged the mugs on a table at a craft fair. The arrangement is shown viewed from above. Draw two *different* ways that the mugs can be arranged in rows so that each row has the same number of mugs.

Chapter 3 Performance Task Student Model 2

NAME _____ DATE _____ SCORE _____

Performance Task

Crafty Artists

Students in the Future Artists club decorate and sell t-shirts and mugs. The table shows the supplies that the club ordered during November.

Item	Number Ordered	Cost per Item
plain t-shirt	40	$3
plain mug	24	$2
box of fabric markers	24	$10
box of paint markers	36	$12

Write your answers on another piece of paper. Show all your work to receive full credit.

Part A

The club orders an *additional* 2 cartons of fabric markers. Each carton contains 8 boxes of markers. To find the total cost, Ian first calculated 2 × 8. He then multiplied the result by the cost of each box. Mila also calculated the total. She first found the cost of 8 boxes of markers. She then multiplied the result by the number of cartons. Write a single number statement that shows that Ian and Mila's totals are equal. Explain the property that the number statement demonstrates.

2×8×10=8×10×2   Commutative Property of Mult.

Part B

Write a word statement that compares the cost of a box of paint markers to the cost of a plain t-shirt. Then write a number statement involving multiplication that shows the same comparison.

A box of paint markers costs 4 times as much as a t-shirt. 12=4×3

Performance Task (continued)

Part C

Plain t-shirts come in packages of 4. In order to find p, the number of packages of t-shirts in the order, Jada writes the equation $40 \div 4 = p$. Write two *different* equations, one using multiplication, and the other using division, that could also be used to find p. Then find the value of p.

$p \times 4 = 40$
$p = 10$

Part D

The club ordered three times as many boxes of paint markers in November as it did in September. Use the diagram below to show how you can find the number of boxes of paint markers the club ordered in September. Then state the number of boxes ordered in September.

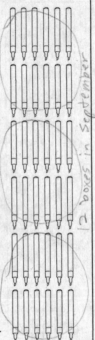

12 boxes in September

Part E

The club members decorated all of the mugs ordered in November. They arranged the mugs on a table at a craft fair. The arrangement is shown viewed from above. Draw two *different* ways that the mugs can be arranged in rows so that each row has the same number of mugs.

Chapter 3 Performance Task Student Model 3

NAME _____ DATE _____

SCORE _____

Performance Task

Crafty Artists

Students in the Future Artists club decorate and sell t-shirts and mugs. The table shows the supplies that the club ordered during November.

Item	Number Ordered	Cost per Item
plain t-shirt	40	$3
plain mug	24	$2
box of fabric markers	24	$10
box of paint markers	36	$12

Write your answers on another piece of paper. Show all your work to receive full credit.

Part A

The club orders an *additional* 2 cartons of fabric markers. Each carton contains 8 boxes of markers. To find the total cost, Ian first calculated 2 × 8. He then multiplied the result by the cost of each box. Mila also calculated the total. She first found the cost of 8 boxes of markers. She then multiplied the result by the number of cartons. Write a single number statement that shows that Ian and Mila's totals are equal. Explain the property that the number statement demonstrates.

$(2 \times 8) \times 10 = 2 \times (8 \times 10)$

Part B

Write a word statement that compares the cost of a box of paint markers to the cost of a plain t-shirt. Then write a number statement involving multiplication that shows the same comparison.

$12 = 4 \times 3$

Performance Task (continued)

Part C

Plain t-shirts come in packages of 4. In order to find p, the number of packages of t-shirts in the order, Jada writes the equation $40 \div 4 = p$. Write two *different* equations, one using multiplication, and the other using division, that could also be used to find p. Then find the value of p.

$p \cdot 40 = 4$ $p \times 4 = 40$ $p = 10$

Part D

The club ordered three times as many boxes of paint markers in November as it did in September. Use the diagram below to show how you can find the number of boxes of paint markers the club ordered in September. Then state the number of boxes ordered in September.

$36 \div 3 = 12$ boxes of markers.

Part E

The club members decorated all of the mugs ordered in November. They arranged the mugs on a table at a craft fair. The arrangement is shown viewed from above. Draw two *different* ways that the mugs can be arranged in rows so that each row has the same number of mugs.

Chapter 3 Performance Task Student Model 4

NAME _____ DATE _____

SCORE _____

Performance Task

Crafty Artists

Students in the Future Artists club decorate and sell t-shirts and mugs. The table shows the supplies that the club ordered during November.

Item	Number Ordered	Cost per Item
plain t-shirt	40	$3
plain mug	24	$2
box of fabric markers	24	$10
box of paint markers	36	$12

Write your answers on another piece of paper. Show all your work to receive full credit.

Part A

The club orders an *additional* 2 cartons of fabric markers. Each carton contains 8 boxes of markers. To find the total cost, Ian first calculated 2×8. He then multiplied the result by the cost of each box. Mila also calculated the total. She first found the cost of 8 boxes of markers. She then multiplied the result by the number of cartons. Write a single number statement that shows that Ian and Mila's totals are equal. Explain the property that the number statement demonstrates.

$2 \times (8 \times 10) = (2 \times 8) \times 10$

Part B

Write a word statement that compares the cost of a box of paint markers to the cost of a plain t-shirt. Then write a number statement involving multiplication that shows the same comparison.

A box of paint markers costs $9 more than a plain t-shirt.
$12 = 3 + 9$

Performance Task (continued)

Part C

Plain t-shirts come in packages of 4. In order to find p, the number of packages of t-shirts in the order, Jada writes the equation $40 \div 4 = p$. Write two *different* equations, one using multiplication, and the other using division, that could also be used to find p. Then find the value of p.

$p = 10$

Part D

The club ordered three times as many boxes of paint markers in November as it did in September. Use the diagram below to show how you can find the number of boxes of paint markers the club ordered in September. Then state the number of boxes ordered in September.

$36 \times 3 = 108$ They ordered 108 boxes in September.

Part E

The club members decorated all of the mugs ordered in November. They arranged the mugs on a table at a craft fair. The arrangement is shown viewed from above. Draw two *different* ways that the mugs can be arranged in rows so that each row has the same number of mugs.

Chapter 4 Performance Task Rubric

Page 143 • Ocean Camp

Task Scenario Students will use knowledge of place value, multiplication skills, and understanding of reasonableness to find and estimate quantities that arise in situations involving marine life and camp activities.		
Depth of Knowledge		DOK2, DOK3
Part	**Maximum Points**	**Scoring Rubric**
A	2	Full Credit: $$\begin{array}{r} \overset{1\;4}{428} \\ \times\;\;\;\;6 \\ \hline 2{,}568 \end{array}$$ Partial Credit (1 point) will be given for one error (such as one instance of not regrouping). No credit will be given if there is no work shown **OR** if the answer is incorrect AND the work does not support the answer.
B	2	Full Credit: No; 62 is close to 60, and 60 × 6 = 360. So, Lily would not be able to drive 428 miles in 6 hours if she averages 62 miles per hour. Partial Credit (1 point) will be given if the exact product is given, and the answer is correct, but the student does not use estimation. No credit will be given if there is no explanation/work shown.
C	3	Full Credit: I have to multiply 12,000 by 5. To do this, I can multiply 12 by 5, which is 60. Because 12,000 = 12 × 1,000, I can then put 3 zeros on the end of 60 to show multiplying by 1,000. So, a gray whale migrates about 60,000 miles along the coast in 5 years. Partial Credit (2 points) will be given if the student shows multiplication by 24,000 instead of 12,000 **OR** if the student does not indicate how to get from 60 to 60,000. Partial Credit (1 point) will be given for a correct answer with direct multiplication shown but no indication of multiplying 12 by 5 (2-digit by 1-digit). No credit will be given if the answer is correct but no work is shown.

Chapter 4 Performance Task Rubric, continued

Part	Maximum Points	Scoring Rubric
D	4	Full Credit: [area model: 50	2 with height 7] 50 × 7 = 350 2 × 7 = 14 350 + 14 = 364 pounds of fish Partial Credit (3 points) will be given if there is one error/missing product in the area model **OR** one error/missing equation in the partial and final products. Partial Credit (2 points) will be given if there are two errors/missing products in the area model and/or partial and final products. Partial Credit (1 point) will be given if there are three errors/missing products in the area model and/or partial and final products. No credit will be given if there are no area model and no partial products shown, even if the answer is correct.
E	3	Full Credit: Bryan plus the 15 other campers and 2 leaders makes 18 people. 18 rounds to 20, and 53 rounds to 50. 20 × 50 = 1,000. So, the trailer would have to be able to haul around 1,000 pounds. Trailers A, D, and E will work because each can haul 1,050 or more pounds. Trailer B, which can haul 600 pounds, and trailer C, which can haul 900 pounds, will not work. Partial Credit (2 points) will be given if the rounding and product are correct but the correct trailers are not identified. Partial Credit (1 point) will be given if the student only rounds one of the numbers (18 or 53) to find the estimate **OR** finds the exact value instead of estimating. No credit will be given for an answer/work that is completely incorrect **OR** if there is no explanation/work shown.
TOTAL	14	

Chapter 4 Performance Task Student Model 1

NAME _____ DATE _____ SCORE _____

Performance Task

Ocean Camp

This summer, Bryan is spending a month at an ocean camp that is 428 miles from his home. He is studying ocean wildlife and participating in recreational activities.

Write your answers on another piece of paper. Show all your work to receive full credit.

Part A

Bryan's parents drove him to the camp in the family car, and then returned home. They will drive the car the same route to pick Bryan up at the end of the month. Bryan's older sister Lily plans to drive the family car to the camp in the middle of the month to visit him, and then return home. How many total miles will the car be driven to and from the camp this summer? Show your work.

428
× 6
2568

Part B

Lily says that she can usually drive 62 miles per hour on the highway that leads to the camp. Is it reasonable to assume that Lily can make it to the camp in 6 hours if she drives at that rate? Explain your answer using estimation. No. 62 rounds to 60, and 60×6=360, which is much less than 428.

Part C

Bryan is studying gray whales, which migrate north and south along the Pacific coast of North America each year. The round is trip approximately 12,000 miles. Explain how you can use multiplication of a one-digit number by a two-digit number to find the approximate number of miles a gray whale migrates along the Pacific coast in 5 years.

12,000×5 = 12×1000×5 = 12×5×1000
= 60×1000
= 60,000 miles

Performance Task (continued)

Part D

Bryan volunteers for one full week feeding sea lion pups at the marine sanctuary. There are 4 pups that each eat 13 pounds of fish per day. Complete the area model to show the total number of pounds of fish that Bryan feeds the sea lion pups for one week. Then write the partial products and the total product.

Each day
4 × 13
52

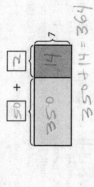

350 + 14 = 364 pounds

Part E

Bryan is going on an all-day sea kayaking adventure with 15 other campers and two camp leaders. Each person will get a kayak. Each kayak weighs 53 pounds. The table shows the maximum weights that 5 different kayak trailers can haul. The group will use one kayak trailer. Use estimation to figure out which trailers can haul all of the kayaks needed for the adventure. Explain your answer.

Trailer	A	B	C	D	E
Maximum Weight (pounds)	1,250	600	900	1,500	1,050

18 people
53 pounds
Round 18 to 20, and 53 to 50. 20×50=1000 pounds.
Trailers A, D, and E can haul at least 1000 pounds.

Chapter 4 Performance Task Student Model 2

NAME _____ DATE _____ SCORE _____

Performance Task

Ocean Camp

This summer, Bryan is spending a month at an ocean camp that is 428 miles from his home. He is studying ocean wildlife and participating in recreational activities.

Write your answers on another piece of paper. Show all your work to receive full credit.

Part A

Bryan's parents drove him to the camp in the family car, and then returned home. They will drive the car the same route to pick Bryan up at the end of the month. Bryan's older sister Lily plans to drive the family car to the camp in the middle of the month to visit him, and then return home. How many total miles will the car be driven to and from the camp this summer? Show your work.

Round 62 to 60. 6×60 = 360
This is a lot less than 428.

Part B

Lily says that she can usually drive 62 miles per hour on the highway that leads to the camp. Is it reasonable to assume that Lily can make it to the camp in 6 hours if she drives at that rate? Explain your answer using estimation.

Part C

Bryan is studying gray whales, which migrate north and south along the Pacific coast of North America each year. The round is trip approximately 12,000 miles. Explain how you can use multiplication of a one-digit number by a two-digit number to find the approximate number of miles a gray whale migrates along the Pacific coast in 5 years.

5 × (12 × 1000) = (5×12) × 1000
= 60,000 miles

428
× 6
2468

Performance Task (continued)

Part D

Bryan volunteers for one full week feeding sea lion pups at the marine sanctuary. There are 4 pups that each eat 13 pounds of fish per day. Complete the area model to show the total number of pounds of fish that Bryan feeds the sea lion pups for one week. Then write the partial products and the total product.

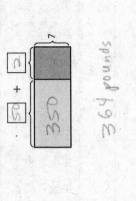

364 pounds

Part E

Bryan is going on an all-day sea kayaking adventure with 15 other campers and two camp leaders. Each person will get a kayak. Each kayak weighs 53 pounds. The table shows the maximum weights that 5 different kayak trailers can haul. The group will use one kayak trailer. Use estimation to figure out which trailers can haul all of the kayaks needed for the adventure. Explain your answer.

Trailer	A	B	C	D	E
Maximum Weight (pounds)	1,250	600	900	1,500	1,050

(D and E circled)

18 × 53 (people × weight per kayak)
round ↓
20 × 50 = 1000

Chapter 4 Performance Task Student Model 3

NAME _____ DATE _____

SCORE _____

Performance Task

Ocean Camp

This summer, Bryan is spending a month at an ocean camp that is 428 miles from his home. He is studying ocean wildlife and participating in recreational activities.

Write your answers on another piece of paper. Show all your work to receive full credit.

Part A

Bryan's parents drove him to the camp in the family car, and then returned home. They will drive the car the same route to pick Bryan up at the end of the month. Bryan's older sister Lily plans to drive the family car to the camp in the middle of the month to visit him, and then return home. How many total miles will the car be driven to and from the camp this summer? Show your work.

428
× 6
2468

2,468

Part B

Lily says that she can usually drive 62 miles per hour on the highway that leads to the camp. Is it reasonable to assume that Lily can make it to the camp in 6 hours if she drives at that rate? Explain your answer using estimation.

62
× 6
372

372 < 428

Part C

Bryan is studying gray whales, which migrate north and south along the Pacific coast of North America each year. The round is trip approximately 12,000 miles. Explain how you can use multiplication of a one-digit number by a two-digit number to find the approximate number of miles a gray whale migrates along the Pacific coast in 5 years. 60,000 miles

Performance Task (continued)

Part D

Bryan volunteers for one full week feeding sea lion pups at the marine sanctuary. There are 4 pups that each eat 13 pounds of fish per day. Complete the area model to show the total number of pounds of fish that Bryan feeds the sea lion pups for one week. Then write the partial products and the total product.

1
52
× 7
364 pounds

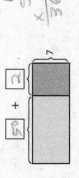

50 + 2 } 7

Part E

Bryan is going on an all-day sea kayaking adventure with 15 other campers and two camp leaders. Each person will get a kayak. Each kayak weighs 53 pounds. The table shows the maximum weights that 5 different kayak trailers can haul. The group will use one kayak trailer. Use estimation to figure out which trailers can haul all of the kayaks needed for the adventure. Explain your answer.

Trailer	A	B	C	D	E
Maximum Weight (pounds)	1,250	600	900	1,500	1,050

Round 53 to 50.

15
× 50
750

All trailers except B will work

Chapter 4 Performance Task Student Model 4

NAME _____ DATE _____

SCORE _____

Performance Task

Ocean Camp

This summer, Bryan is spending a month at an ocean camp that is 428 miles from his home. He is studying ocean wildlife and participating in recreational activities.

Write your answers on another piece of paper. Show all your work to receive full credit.

Part A

Bryan's parents drove him to the camp in the family car, and then returned home. They will drive the car the same route to pick Bryan up at the end of the month. Bryan's older sister Lily plans to drive the family car to the camp in the middle of the month to visit him, and then return home. How many total miles will the car be driven to and from the camp this summer? Show your work. **1,284 miles**

428
× 3
1,284

Part B

Lily says that she can usually drive 62 miles per hour on the highway that leads to the camp. Is it reasonable to assume that Lily can make it to the camp in 6 hours if she drives at that rate? Explain your answer using estimation. **No, she can't do it. 372 < 428**

62
× 6
372

Part C

Bryan is studying gray whales, which migrate north and south along the Pacific coast of North America each year. The round is trip approximately 12,000 miles. Explain how you can use multiplication of a one-digit number by a two-digit number to find the approximate number of miles a gray whale migrates along the Pacific coast in 5 years. **12 × 5 = 60**

Performance Task (continued)

Part D

Bryan volunteers for one full week feeding sea lion pups at the marine sanctuary. There are 4 pups that each eat 13 pounds of fish per day. Complete the area model to show the total number of pounds of fish that Bryan feeds the sea lion pups for one week. Then write the partial products and the total product.

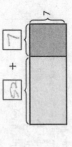

Part E

Bryan is going on an all-day sea kayaking adventure with 15 other campers and two camp leaders. Each person will get a kayak. Each kayak weighs 53 pounds. The table shows the maximum weights that 5 different kayak trailers can haul. The group will use one kayak trailer. Use estimation to figure out which trailers can haul all of the kayaks needed for the adventure. Explain your answer.

Trailer	A	B	C	D	E
Maximum Weight (pounds)	1,250	600	900	1,500	1,050

(C, D, E circled)

50
× 15
250
500
750

Chapter 5 Performance Task Rubric

Page 145 • Happy Paws

Task Scenario
Students will use knowledge of place value and two-digit multiplication to model, calculate, and compare products and estimates of products using quantities involved in pet care activities.

Depth of Knowledge	DOK2, DOK3

Part	Maximum Points	Scoring Rubric

A	4	Full Credit:
		
		   10 + 2
		  85 { [850] [170] }
		
		Number of Pounds of Dog Food: $850 + 170 = 1{,}020$
		Partial Credit (3 points) will be given if there is one mistake in the partial product calculations and the total is consistent with the mistake.
		Partial Credit (2 points) will be given if there are two mistakes in the partial product calculations and the total is consistent with the mistake.
		Partial Credit (1 point) will be given if there are 3 mistakes in the partial product calculations and the total is consistent with the mistake.
		No credit will be given for a correct total with no area model shown.

B	3	Full Credit: There are 10 more pounds of cat food than dog food donated each month. $10 \times 12 = 120$. $120 + 1{,}020 = 1{,}140$ pounds of cat food. (Give full credit if the student correctly uses an incorrect total from **Part A**.) Partial Credit (2 points) will be given if any one of the explanation, multiplication work, or addition work is incorrect or missing. Partial Credit (1 point) will be given if any two of the explanation, multiplication work, or addition work are incorrect or missing AND/OR if the student finds the total by multiplying 95 by 12. No credit will be given for a correct total with no work or explanation.

Chapter 5 Performance Task Rubric, continued

Part	Maximum Points	Scoring Rubric
C	2	Full Credit: Jen did not put a 0 in the ones place for the second row of multiplication to show that she was multiplying by 2 tens instead of 2 ones. $$\begin{array}{r} 53 \\ \times\ 25 \\ \hline 265 \\ +\ 1060 \\ \hline 1325 \end{array}$$ Partial Credit (1 point) will be given if either the explanation or correction is missing or incorrect. No credit will be given if there is no explanation or correction.
D	2	Full Credit: $s = 40 \times 20$ $s = 800$ square feet Partial Credit (1 point) will be given if the student multiplies incorrectly. No credit will be given if the answer is correct but no work is shown **OR** if the student rounds incorrectly **OR** gives an exact number instead of an estimate.
E	2	Full Credit: The estimate is greater because I rounded both 38 and 16 up. It is better for the estimate to be greater because it would be OK for each cat to have more space but not OK if the cats do not have enough space. Partial Credit (1 point) will be given if one explanation is missing or incorrect. No credit will be given for a "greater than" answer that does not include an explanation.
TOTAL	13	

Chapter 5 Performance Task Student Model 1

Performance Task

Happy Paws

Jen and Quon are helping out Happy Paws animal rescue for their school service project.

Write your answers on another piece of paper. Show all your work to receive full credit.

Part A

Jen and Quon asked a pet food company for food donations. The company agreed to donate 85 pounds of dog food each month for a year. Complete the area model for the total number of pounds of dog food that the company will donate in one year. Then use the model to find the total.

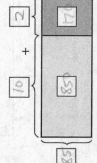

Number of Pounds of Dog Food: $850 + 170 = 1,020$

Part B

The company also agreed to donate 95 pounds of cat food to Happy Paws each month for a year. Explain how to use your answer from Part A to find the total number of pounds of cat food that will be donated in one year. Then state the total.

There are 10 more pounds of cat food than dog food donated each month.
$10 \times 12 = 120$
$1,020 + 120 = 1,140$ pounds of cat food

Performance Task (continued)

Part C

Happy Paws has 53 dogs in foster homes. The director has asked Jen to pack a box of canned dog food for each foster home. Each box will contain 25 cans. Jen does the multiplication shown to figure out how many cans she will need. Use place value to explain the error that Jen made. Then correct the error and find the total number of cans.

```
   53
 × 25
  265
+106
  371
```

Jen did not put a 0 in the ones place for the second row of the multiplication.

```
   53
 × 25
  265
+1060
 1325
```

Part D

Quon is helping set up an outdoor cat pen for the 38 cats that live at Happy Paws. Each cat should have at least 16 square feet of space. Write an equation that can be used to find s, an estimate of the total number of square feet the pen should enclose. Then solve the equation.

$s = 40 \times 20$
$s = 800$ square feet

Part E

Is your estimate from Part D greater than or less than the actual number of square feet? Explain. In this case, explain whether it better for the estimate to be less than or greater than the actual number.

Greater, because I rounded 38 and 16 up. It is better to be greater because it is better for each cat to have more space than required than to have less than required.

Chapter 5 Performance Task Student Model 2

Performance Task

Happy Paws

Jen and Quon are helping out Happy Paws animal rescue for their school service project.

Write your answers on another piece of paper. Show all your work to receive full credit.

Part A

Jen and Quon asked a pet food company for food donations. The company agreed to donate 85 pounds of dog food each month for a year. Complete the area model for the total number of pounds of dog food that the company will donate in one year. Then use the model to find the total.

	10	+	2
85	850		170

Number of Pounds of Dog Food: 920

Part B

The company also agreed to donate 95 pounds of cat food to Happy Paws each month for a year. Explain how to use your answer from Part A to find the total number of pounds of cat food that will be donated in one year. Then state the total.

```
   95
 ×  12
   190
 + 950
  1,140
```
1,140 pounds of cat food

Performance Task (continued)

Part C

Happy Paws has 53 dogs in foster homes. The director has asked Jen to pack a box of canned dog food for each foster home. Each box will contain 25 cans. Jen does the multiplication shown to figure out how many cans she will need. Use place value to explain the error that Jen made. Then correct the error and find the total number of cans.

```
   53
 × 25
  265
+ 106
  371
```

She forgot to put a 0 in the second row.

```
   53
 × 25
  265
+1060
 1,325
```

Part D

Quon is helping set up an outdoor cat pen for the 38 cats that live at Happy Paws. Each cat should have at least 16 square feet of space. Write an equation that can be used to find s, an estimate of the total number of square feet the pen should enclose. Then solve the equation.

$s = 40 \times 20$
$s = 800$

Part E

Is your estimate from Part D greater than or less than the actual number of square feet? Explain. In this case, explain whether it better for the estimate to be less than or greater than the actual number.

Greater. Both numbers were rounded up before multiplying. This gives each cat more space.

Chapter 5 Performance Task Student Model 3

NAME _____ **DATE** _____ **SCORE** _____

Performance Task

Happy Paws

Jen and Quon are helping out Happy Paws animal rescue for their school service project.

Write your answers on another piece of paper. Show all your work to receive full credit.

Part A

Jen and Quon asked a pet food company for food donations. The company agreed to donate 85 pounds of dog food each month for a year. Complete the area model for the total number of pounds of dog food that the company will donate in one year. Then use the model to find the total.

```
     85
   x 12
   ----
    170
  + 850
  -----
   1020
```

[Area model: 12 | + | 80 | 5 ; bottom shows 85]

Number of Pounds of Dog Food: 1020

Part B

The company also agreed to donate 95 pounds of cat food to Happy Paws each month for a year. Explain how to use your answer from Part A to find the total number of pounds of cat food that will be donated in one year. Then state the total.

```
    95
  x 12
  ----
   190
 + 950
 -----
  1140  pounds
```

Performance Task (continued)

Part C

Happy Paws has 53 dogs in foster homes. The director has asked Jen to pack a box of canned dog food for each foster home. Each box will contain 25 cans. Jen does the multiplication shown to figure out how many cans she will need. Use place value to explain the error that Jen made. Then correct the error and find the total number of cans.

```
     53
   x 25
   ----
    265
  + 106
    371
```

Jen didn't line the numbers up right.

```
    265
 + 1060
  -----
   1325
```

Part D

Quon is helping set up an outdoor cat pen for the 38 cats that live at Happy Paws. Each cat should have at least 16 square feet of space. Write an equation that can be used to find s, an estimate of the total number of square feet the pen should enclose. Then solve the equation.

$s = 40 \times 20 = 800$

Part E

Is your estimate from Part D greater than or less than the actual number of square feet? Explain. In this case, explain whether it better for the estimate to be less than or greater than the actual number.

Greater because I rounded both 38 and 16 up.

Chapter 5 Performance Task Student Model 4

NAME _____ **DATE** _____

SCORE _____

Performance Task

Happy Paws

Jen and Quon are helping out Happy Paws animal rescue for their school service project.

Write your answers on another piece of paper. Show all your work to receive full credit.

Part A

Jen and Quon asked a pet food company for food donations. The company agreed to donate 85 pounds of dog food each month for a year. Complete the area model for the total number of pounds of dog food that the company will donate in one year. Then use the model to find the total.

```
  85
 x 12
 ----
  170
+ 850
-----
 1020
```

Number of Pounds of Dog Food: 1,020

Part B

The company also agreed to donate 95 pounds of cat food to Happy Paws each month for a year. Explain how to use your answer from Part A to find the total number of pounds of cat food that will be donated in one year. Then state the total.

```
   95
  x 12
  ----
   190
 + 950
 -----
  1140
```

Performance Task (continued)

Part C

Happy Paws has 53 dogs in foster homes. The director has asked Jen to pack a box of canned dog food for each foster home. Each box will contain 25 cans. Jen does the multiplication shown to figure out how many cans she will need. Use place value to explain the error that Jen made. Then correct the error and find the total number of cans.

```
    53
   x25
  2650
 +1060
 -----
  3710
```

Jen didn't line the numbers up right.

```
    53
   x 25
   ---
   265
 + 106
 -----
   371
```

Part D

Quon is helping set up an outdoor cat pen for the 38 cats that live at Happy Paws. Each cat should have at least 16 square feet of space. Write an equation that can be used to find s, an estimate of the total number of square feet the pen should enclose. Then solve the equation.

```
  38
 x16
 ---
 228
+380
----
 608
```

$s = 38 \times 16$
$s = 608$

Part E

Is your estimate from Part D greater than or less than the actual number of square feet? Explain. In this case, explain whether it better for the estimate to be less than or greater than the actual number.

It's better for the estimate to be greater.

Chapter 6   Performance Task Rubric

Page 147 • Hiking the Pacific Crest Trail

Task Scenario

Students will extend, describe, and analyze patterns and sequences found in displays of grocery store items. Students will need to perform research in advance. They will need to research the length of the Pacific Crest Trail, the height of Mt. Hood, and the number of major mountain passes along the Pacific Crest Trail.

Depth of Knowledge	DOK2, DOK3	
Part	**Maximum Points**	**Scoring Rubric**
A	3	Full Credit: The length of the trail is 2,650, and I need to divide by 9 months. Using compatible numbers, I rounded 2,650 to 2,700. 2,700 ÷ 9 = 300. They will hike about 300 miles each month. This is greater than the number of miles they would actually hike each month because I rounded up. Partial Credit (2 points) will be given if one of the estimate, work, or greater than/less than comparison is incorrect or missing. Partial Credit (1 point) will be given if two of the estimate, work, or greater than/less than comparison are incorrect or missing. No credit will be given if all parts are incorrect or if only the estimate of 300 is given with no work shown and no answer to the greater than/less than question.
B	3	Full Credit: I rounded 2,650 to 3,000 and 9 to 10. 3,000 ÷ 10 = 300. This is the same answer that I got by using compatible numbers because in both cases, I rounded 2,650 up, but 2,700 is a little less than 3,000 just like 9 is a little less than 10. Partial Credit (2 points) will be given if one of the estimate, work, or comparison with the answer from **Part A** is incorrect or missing. Partial Credit (1 point) will be given if two of the estimate, work, or comparison with the answer from **Part A** are incorrect or missing. No credit will be given if all parts are incorrect or if only the estimate of 300 is given with no work shown and there is no comparison with **Part A**.
C	2	Full Credit: 11,250 − 8,500 = 2,750 2,750 ÷ 10 = 275 feet per hour. Partial Credit (1 point) will be given for a correct answer with incorrect work **OR** for correct work but incorrect answer. No credit will be given if the answer is correct but no work is shown.

Chapter 6 Performance Task Rubric, continued

Part	Maximum Points	Scoring Rubric
D	2	Full Credit: 13 pounds $$\begin{array}{r} 13 \\ 6\overline{)78} \\ -\,6 \\ \hline 18 \\ -18 \\ \hline 0 \end{array}$$ Partial Credit (1 point) will be given for a correct answer with incorrect work **OR** for correct work but incorrect answer. No credit will be given if the answer is correct but no work is shown.
E	3	Full Credit: No. There are 57 major mountain passes along the trail. If I divide 57 by 6, I get a remainder of 3. A remainder means that 57 photos cannot be divided equally among 6 people. $$\begin{array}{r} 9R3 \\ 6\overline{)57} \\ -54 \\ \hline 3 \end{array}$$ Partial Credit (2 points) will be given if the work is wrong and the answer (yes/no) matches the work **OR** if the answer and explanation are correct but the student does not show/explain how s/he knows that there is a remainder. Partial Credit (1 point) will be given if the division is correct but the answer is wrong **OR** if the answer is correct and there is an explanation, but remainders are not mentioned. No credit will be given for a yes/no answer that does not include an explanation or any work.
TOTAL	13	

Performance Task

Hiking the Pacific Crest Trail

Rodrigo plans to section hike the length of the Pacific Crest Trail with 5 friends. The trail runs from Mexico to Canada through California, Washington, and Oregon.

Write your answers on another piece of paper. Show all your work to receive full credit.

Part A

The friends will hike through sections of the trail during 3 summers, over a total of 9 months. They want to cover about the same distance each month. Research the length of the trail in miles. Use compatible numbers to estimate the number of miles they will hike each month. State whether your estimate is greater than or less than the actual number of miles they would hike each month.

length = 2650 ÷ 9
2700 ÷ 9 = 300 miles per month
This is greater than the actual number of miles.

Part B

Now estimate the distance the friends will hike each month by rounding the numbers to the largest possible place value. Is this distance less than, equal to, or greater than your estimate from Part A? Why do you think this is so?

2,650 → 3,000
9 → 10
3000 ÷ 10 = 300; the same as Part A.
I rounded up all numbers in both parts.

Performance Task (continued)

Part C

Along their route, the hikers plan to climb to the top of Mt. Hood in Oregon. They will allow 10 hours to reach the summit. Research the height of Mt. Hood in feet. Use the closest multiple of 10 for the height. To reach the summit, how many feet in elevation must they gain each hour from their base camp of 8,500 feet?

11,250
−8,500
———
2,750

2,750 ÷ 10 = 275 feet per hour

Part D

The hikers will carry backpacks with their own personal gear as well as group items whose weight will be equally shared among them. The shared items weigh a total of 78 pounds. Write an equation that can be used to find w, the weight of the shared items carried by each hiker. Then solve the equation.

78 ÷ 6 = w w = 13 pounds

Part E

Rodrigo and his friends plan to take one group photo every time they go through a major mountain pass along the trail. Research the number of major mountain passes along the Pacific Crest Trail. Is it possible for each of them to take the same number of such photos? Use your knowledge of remainders to explain why or why not.

No, there are 57 major mountain passes. 57 cannot be divided evenly by 6.

  9 r3
6)57
 −54
 ———
   3

Chapter 6 Performance Task Student Model 2

Performance Task

Hiking the Pacific Crest Trail

Rodrigo plans to section hike the length of the Pacific Crest Trail with 5 friends. The trail runs from Mexico to Canada through California, Washington, and Oregon.

Write your answers on another piece of paper. Show all your work to receive full credit.

Part A

The friends will hike through sections of the trail during 3 summers, over a total of 9 months. They want to cover about the same distance each month. Research the length of the trail in miles. Use compatible numbers to estimate the number of miles they will hike each month. State whether your estimate is greater than or less than the actual number of miles they would hike each month.

$2700 \div 9 = 300$ miles per month

The estimate is greater than the actual.

Part B

Now estimate the distance the friends will hike each month by rounding the numbers to the largest possible place value. Is this distance less than, equal to, or greater than your estimate from Part A? Why do you think this is so?

$2650 \div 9$

$3000 \div 10 = 300$

Performance Task (continued)

Part C

Along their route, the hikers plan to climb to the top of Mt. Hood in Oregon. They will allow 10 hours to reach the summit. Research the height of Mt. Hood in feet. Use the closest multiple of 10 for the height. To reach the summit, how many feet in elevation must they gain each hour from their base camp of 8,500 feet?

$11,250 \div 10 = 1,125$ feet per hour

Part D

The hikers will carry backpacks with their own personal gear as well as group items whose weight will be equally shared among them. The shared items weigh a total of 78 pounds. Write an equation that can be used to find w, the weight of the shared items carried by each hiker. Then solve the equation.

$78 \div 6 = w$

$w = 13$ pounds

Part E

Rodrigo and his friends plan to take one group photo every time they go through a major mountain pass along the trail. Research the number of major mountain passes along the Pacific Crest Trail. Is it possible for each of them to take the same number of such photos? Use your knowledge of remainders to explain why or why not.

No, there are 57 mountain passes, and $57 \div 6$ has a remainder.

Chapter 6 Performance Task Student Model 3

NAME _____ DATE _____ SCORE _____

Performance Task

Hiking the Pacific Crest Trail

Rodrigo plans to section hike the length of the Pacific Crest Trail with 5 friends. The trail runs from Mexico to Canada through California, Washington, and Oregon.

Write your answers on another piece of paper. Show all your work to receive full credit.

Part A

The friends will hike through sections of the trail during 3 summers, over a total of 9 months. They want to cover about the same distance each month. Research the length of the trail in miles. Use compatible numbers to estimate the number of miles they will hike each month. State whether your estimate is greater than or less than the actual number of miles they would hike each month.

2,700 ÷ 9 = 300 miles per month

Part B

Now estimate the distance the friends will hike each month by rounding the numbers to the largest possible place value. Is this distance less than, equal to, or greater than your estimate from Part A? Why do you think this is so?

300 miles per month. It's the same because I rounded up both times.

Performance Task (continued)

Part C

Along their route, the hikers plan to climb to the top of Mt. Hood in Oregon. They will allow 10 hours to reach the summit. Research the height of Mt. Hood in feet. Use the closest multiple of 10 for the height. To reach the summit, how many feet in elevation must they gain each hour from their base camp of 8,500 feet?

11,250 ÷ 10
1,125 feet per hour

Part D

The hikers will carry backpacks with their own personal gear as well as group items whose weight will be equally shared among them. The shared items weigh a total of 78 pounds. Write an equation that can be used to find w, the weight of the shared items carried by each hiker. Then solve the equation.

6)78
13 pounds

Part E

Rodrigo and his friends plan to take one group photo every time they go through a major mountain pass along the trail. Research the number of major mountain passes along the Pacific Crest Trail. Is it possible for each of them to take the same number of such photos? Use your knowledge of remainders to explain why or why not.

No. The 57 major mountain passes can't be evenly divided by 6.

Chapter 6 Performance Task Student Model 4

Performance Task

Hiking the Pacific Crest Trail

Rodrigo plans to section hike the length of the Pacific Crest Trail with 5 friends. The trail runs from Mexico to Canada through California, Washington, and Oregon.

Write your answers on another piece of paper. Show all your work to receive full credit.

Part A

The friends will hike through sections of the trail during 3 summers, over a total of 9 months. They want to cover about the same distance each month. Research the length of the trail in miles. Use compatible numbers to estimate the number of miles they will hike each month. State whether your estimate is greater than or less than the actual number of miles they would hike each month.

2700 ÷ 9 = 300 miles per month

Part B

Now estimate the distance the friends will hike each month by rounding the numbers to the largest possible place value. Is this distance less than, equal to, or greater than your estimate from Part A? Why do you think this is so?

300 miles per month

Performance Task (continued)

Part C

Along their route, the hikers plan to climb to the top of Mt. Hood in Oregon. They will allow 10 hours to reach the summit. Research the height of Mt. Hood in feet. Use the closest multiple of 10 for the height. To reach the summit, how many feet in elevation must they gain each hour from their base camp of 8,500 feet?

11,250
− 8,500
———
 2,750 feet

Part D

The hikers will carry backpacks with their own personal gear as well as group items whose weight will be equally shared among them. The shared items weigh a total of 78 pounds. Write an equation that can be used to find w, the weight of the shared items carried by each hiker. Then solve the equation.

13 pounds

Part E

Rodrigo and his friends plan to take one group photo every time they go through a major mountain pass along the trail. Research the number of major mountain passes along the Pacific Crest Trail. Is it possible for each of them to take the same number of such photos? Use your knowledge of remainders to explain why or why not.

No. there are 57 major mountain passes. 57 can't be evenly divided by 6.

Page 149 • Designing Store Displays

Task Scenario		
Students will extend, describe, and analyze patterns and sequences found in displays of grocery store items.		
Depth of Knowledge		DOK2, DOK3
Part	**Maximum Points**	**Scoring Rubric**
A	2	Full Credit:  Partial Credit (1 point) will be given for one correct row. No credit will be given if both rows are incorrect.
B	2	Full Credit: $c = 10 - r$ **OR** $r + c = 10$ **OR** equivalent. To get the can count for the next row, subtract 1 from the previous row's number of cans. Since there is no a previous row to Row 1, it could be 10. Then $10 - 1 = 9$ for Row 1, which works; $10 - 2 = 8$ for Row 2, which also works, and so on. Partial Credit (1 point) will be given if the equation is correct **OR** if the student correctly states a rule for the pattern but the equation is incorrect or missing. No credit will be given if both the equation and explanation are incorrect.
C	2	Full Credit: $c = 10 - r = 10 - 7 = 3$. There are 3 cans in Row 7. Partial Credit (1 point) will be given for a correct answer with work shown by counting backward from previous rows. No credit will be given if the answer is correct but no work is shown.

Chapter 7 Performance Task Rubric, continued

Part	Maximum Points	Scoring Rubric
D	2	**Full Credit:** There will be 9 rows. There are 5 cans in Row 5, 4 cans in Row 6, 3 cans in Row 7, and 2 cans in Row 8. There is 1 can in Row 9, which has to be the last row because after that, you cannot subtract any more cans. **OR** The last row is the top row and will have 1 can. I substituted 1 into the equation for the number of cans and found that the row number with 1 can is Row 9. Partial Credit (1 point) will be given if either the explanation/methodology is correct **OR** the answer is correct. No credit will be given if the row number is correct but there is no explanation.
E	4	**Full Credit:** | Pile Number	1	2	3	4	5
---	---	---	---	---	---
Number of Lemons	3	5	7	9	11	 Rule: Add 2 **OR** Rule: Odd numbers starting with 3. Partial Credit (3 points) will be given if the table is correct but the rule is incorrect **OR** if 1 number in the table is incorrect. Partial Credit (2 points) will be given if the table has 2 incorrect numbers **OR** the table has 1 incorrect number and the rule is incorrect. Partial Credit (1 point) will be given if the table has 3 incorrect numbers but a correct rule **OR** the table has 2 incorrect numbers and the rule is incorrect. No credit will be given for an answer/work that is completely incorrect or if only one number on the table is correct and the rule is incorrect.
TOTAL	12	

Performance Task

Designing Store Displays

Cearra is helping her parents at the local market that they own.

Write your answers on another piece of paper. Show all your work to receive full credit.

Part A

Cearra makes a display with soup cans. The first three rows of the display are shown. Extend the pattern by drawing two more rows of cans.

Row 3 5 cans
Row 2 6 cans
Row 1 7 cans
Row 2 8 cans
Row 1 9 cans

Part B

Write an equation that is a rule for finding c, the number of cans in row number r of the display. Explain how you figured out the equation.

$r + c = 10$
$c = 10 - r$

the row # and # of cans add up to 10.

Performance Task (continued)

Part C

Use your equation to find the number of cans in Row 7. Show your work.

$c = 10 - 7$
3 cans

Part D

How many total rows will there be when Cearra is finished with the soup can display? Explain how you know.

Row 9 will have 1 can and will be the last row.

Part E

Cearra makes another display with piles of lemons. The first three piles are shown. Complete the table. Then state a rule for the pattern.

Pile Number	1	2	3	4	5
Number of Lemons	3	5	7	9	11

Rule: Add 2 to the previous number of lemons.

Chapter 7 Performance Task Student Model 2

NAME _____ DATE _____ SCORE _____

Performance Task

Designing Store Displays

Cearra is helping her parents at the local market that they own.

Write your answers on another piece of paper. Show all your work to receive full credit.

Part A

Cearra makes a display with soup cans. The first three rows of the display are shown. Extend the pattern by drawing two more rows of cans.

Row 5
Row 4
Row 3
Row 2
Row 1

Part B

Write an equation that is a rule for finding c, the number of cans in row number r of the display. Explain how you figured out the equation.

$c = r - 1$

To get the can count for the next row, I subtracted 1 from the previous row, which is $r-1$ when r is the row number.

Performance Task (continued)

Part C

Use your equation to find the number of cans in Row 7. Show your work.

$c = r - 1$
$c = 7 - 1$
$c = 6$

Part D

How many total rows will there be when Cearra is finished with the soup can display? Explain how you know.

9 rows

There are 9 cans in Row 1, so there will be 1 can in Row 9.

Part E

Cearra makes another display with piles of lemons that form a pattern. The first three piles are shown. Complete the table. Then state a rule for the pattern.

Pile Number	1	2	3	4	5
Number of Lemons	3	5	7	9	11

Rule: Add 2

Performance Task

Designing Store Displays

Cearra is helping her parents at the local market that they own.

Write your answers on another piece of paper. Show all your work to receive full credit.

Part A

Cearra makes a display with soup cans. The first three rows of the display are shown. Extend the pattern by drawing two more rows of cans.

Row 3
Row 2
Row 1

Part B

Write an equation that is a rule for finding c, the number of cans in row number r of the display. Explain how you figured out the equation.

$c = 7 - 1$

Performance Task (continued)

Part C

Use your equation to find the number of cans in Row 7. Show your work.

$c = 7 - 1$
$c = 6$

6 cans

Part D

How many total rows will there be when Cearra is finished with the soup can display? Explain how you know.

There will be 10 rows because the first row has 70 cans.

Part E

Cearra makes another display with piles of lemons that form a pattern. The first three piles are shown. Complete the table. Then state a rule for the pattern.

Pile Number	1	2	3	4	5
Number of Lemons	3	5	7	9	11

Rule: Odd numbers starting with 3

Chapter 7 Performance Task Student Model 4

NAME _____ DATE _____

SCORE _____

Performance Task

Designing Store Displays

Cearra is helping her parents at the local market that they own.

Write your answers on another piece of paper. Show all your work to receive full credit.

Part A

Cearra makes a display with soup cans. The first three rows of the display are shown. Extend the pattern by drawing two more rows of cans.

Row 3
Row 2
Row 1

Part B

Write an equation that is a rule for finding c, the number of cans in row number r of the display. Explain how you figured out the equation.

Subtract 1 from the previous row.

Performance Task (continued)

Part C

Use your equation to find the number of cans in Row 7. Show your work.

1 can

Part D

How many total rows will there be when Cearra is finished with the soup can display? Explain how you know.

10 rows

Part E

Cearra makes another display with piles of lemons that form a pattern. The first three piles are shown. Complete the table. Then state a rule for the pattern.

Pile Number	1	2	3	4	5
Number of Lemons	3	7	9	11	13

Rule:

Chapter 8 Performance Task Rubric

Page 151 • The River Bank Market Center

Task Scenario
Students will use fractions to determine possible layouts for various shops in a shopping and entertainment center and will determine the possible rental charge for the shops.

Depth of Knowledge	DOK2, DOK3

Part	Maximum Points	Scoring Rubric
A	2	**Full Credit:** Order from greatest to least: Green Door $\frac{5}{12}$, Nerd Nation $\frac{7}{30}$, G-Street Apps $\frac{1}{6}$, Dora's Unbelievably Great Cupcakes $\frac{1}{10}$, Noodlehead's $\frac{1}{20}$, Bike Geek Cycles $\frac{1}{30}$ Partial Credit (1 point) will be given for two items out of rank order. **OR** if student lists items in correct order but from least to greatest rather than greatest to least. No credit will be given for more than two items out of rank order.
B	3	**Full Credit:** For any arrangement featuring the correct number of squares for each client: Green Door 25 squares, Nerd Nation 14 squares, G-Street Apps 10 squares, Dora's Unbelievably Great Cupcakes 6 squares, Noodlehead's 3 squares, Bike Geek Cycles 2 squares. [Grid diagram showing: Green Door Theater, Geek, Noodlehead's, G-Street Apps, Nerd Nation Fitness, Dora's Cupcakes] Partial Credit (2 points) will be given for grids that are correctly divided but fail to be labeled correctly. Partial Credit (1 point) will be given if two or fewer items feature the wrong number of squares. No credit will be given for more than two items that cover an incorrect amount of area.

Grade 4 • Chapter 8 Performance Task Rubric

Chapter 8 Performance Task Rubric, continued

Part	Maximum Points	Scoring Rubric
C	2	**Full Credit:** For an arrangement featuring the correct number of squares for each client: Green Door 25 squares, Nerd Nation 14 squares, G-Street Apps 10 squares, Dora's Unbelievably Great Cupcakes 6 squares, Noodlehead's 3 squares, Bike Geek Cycles 2 squares. Partial Credit (1 point) will be given for any diagram showing two or fewer items that feature the wrong number of squares. No credit will be given for more than two items that cover an incorrect amount of area.
D	3	**Full Credit:** For three arrangements: a 15 × 4 grid, a 10 × 6 grid, and a 20 × 3 grid. Partial Credit (2 points) will be given for providing two correct arrangements **OR** (1 point) will be given for providing one correct arrangement. No credit will be given if no correct arrangement is provided.
E	2	**Full Credit:** For the correct table as shown. |	Least	Greatest
Green Door Theater	$21,250	$25,000
G-Street Apps	$8,500	$10,000
Nerd Nation Fitness	$11,900	$14,000
Dora's Cupcakes	$5,100	$6,000
Noodlehead's	$2,550	$3,000
Bike Geek Cycle Shop	$1,700	$2,000 | Partial Credit (1 point) will be given for providing a table with two or fewer errors. No credit will be given for a table that has more than two errors.
TOTAL	12	

Chapter 8 Performance Task Student Model 1

NAME _____ DATE _____ SCORE _____

Performance Task

The River Bank Market Center

River Bank Market is an urban shopping and entertainment center that is still in its planning stage. River Bank has sold parts of its 1000 square meter space to 6 different client groups. Nerd Nation Fitness has purchased $\frac{7}{30}$ of the space for its center. G-Street Apps has an area equal to $\frac{1}{6}$ of the total space. Bike Geek Cycle Shop has $\frac{1}{30}$ of the center's total space for a shop. Dora's Unbelievably Great Cupcakes has $\frac{1}{10}$ of the space. The Green Door Theater has $\frac{5}{12}$ of the total space. Noodlehead's has $\frac{1}{20}$ of the space.

Part A

The first problem that the center is looking to solve is to determine how much renters should pay. Before they can determine an actual price, the planners must be able to rank the renters in order of the amount of space they are renting. List each renter client from greatest to least with respect to the amount of space that they are renting.

Green Door
Nerd Nation
G-street
Doras
Noodleheads
Cycles

Part B

The River Bank planners need to draw up a floor plan of the building to see how each space lines up. So far, the architects have come up with the floor plan shown below that divides the rectangular space into equal-sized parts that will be divided among the clients. Color in the squares to show a way in which the space for each client can be laid out.

[Floor plan grid with labels: Green Door, G-street, Cycles, Noodle, Nerd Nation, Doras]

Performance Task (continued)

Part C

The clients want to see another variation of the floor plan you made. Draw an additional floor plan and color it in to show each client's space.

[Floor plan grid with labels: Nerd Nation, Doras, Cycles, Noodles, Green Door]

Part D

Having gone over the previous plans, the clients now want the architect to come up with as many rectangular plans for the entire complex as possible. How many different rectangles can you draw? Make a sketch of each rectangle showing the individual squares as you drew above. Each rectangle must be at least 3 squares in width.

Part E

The builders have decided to charge the renters between $850 and $1,000 per square unit as shown in the diagram in **Part B**. Determine the least and greatest amount of rent each client will pay.

	Least	Greatest
Green Door Theater	21,250	25,000
G-Street Apps	8,500	10,000
Nerd Nation Fitness	11,900	14,000
Dora's Cupcakes	5,100	6,000
Noodlehead's	2,550	3,000
Bike Geek Cycle Shop	1,700	2,000

Chapter 8 Performance Task Student Model 2

NAME _____ DATE _____ SCORE _____

Performance Task

The River Bank Market Center

River Bank Market is an urban shopping and entertainment center that is still in its planning stage. River Bank has sold parts of its 1000 square meter space to 6 different client groups. Nerd Nation Fitness has purchased $\frac{7}{30}$ of the space for its center. G-Street Apps has an area equal to $\frac{1}{6}$ of the total space. Bike Geek Cycle Shop has $\frac{1}{30}$ of the center's total space for a shop. Dora's Unbelievably Great Cupcakes has $\frac{1}{10}$ of the space. The Green Door Theater has $\frac{5}{12}$ of the total space. Noodlehead's has $\frac{1}{20}$ of the space.

Part A

The first problem that the center is looking to solve is to determine how much renters should pay. Before they can determine an actual price, the planners must be able to rank the renters in order of the amount of space they are renting. List each renter client from greatest to least with respect to the amount of space that they are renting.

Theater, Fitness, Apps, Cupcakes, Noodle, Cycle

Part B

The River Bank planners need to draw up a floor plan of the building to see how each space lines up. So far, the architects have come up with the floor plan shown below that divides the rectangular space into equal-sized parts that will be divided among the clients. Color in the squares to show a way in which the space for each client can be laid out.

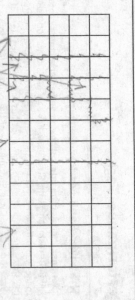

Part C

The clients want to see another variation of the floor plan you made. Draw an additional floor plan and color it in to show each client's space.

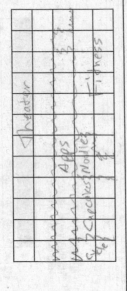

Part D

Having gone over the previous plans, the clients now want the architect to come up with as many rectangular plans for the entire complex as possible. How many different rectangles can you draw? Make a sketch of each rectangle showing the individual squares as you drew above. Each rectangle must be at least 3 squares in width. 3×29, 4×15, 6×10

Part E

The builders have decided to charge the renters between $850 and $1,000 per square unit as shown in the diagram in Part B. Determine the least and greatest amount of rent each client will pay.

	Least	Greatest
Green Door Theater	21,250	25,000
G-Street Apps	8,500	10,000
Nerd Nation Fitness	11,900	14,000
Dora's Cupcakes	5,100	6,000
Noodlehead's	2,550	3,000
Bike Geek Cycle Shop	1,700	2,000

Chapter 8 Performance Task Student Model 3

NAME _____ DATE _____ SCORE _____

Performance Task

The River Bank Market Center

River Bank Market is an urban shopping and entertainment center that is still in its planning stage. River Bank has sold parts of its 1000 square meter space to 6 different client groups. Nerd Nation Fitness has purchased $\frac{7}{30}$ of the space for its center. G-Street Apps has an area equal to $\frac{1}{6}$ of the total space. Bike Geek Cycle Shop has $\frac{5}{30}$ of the center's total space for a shop. Dora's Unbelievably Great Cupcakes has $\frac{1}{10}$ of the space. The Green Door Theater has $\frac{5}{12}$ of the total space. Noodlehead's has $\frac{1}{20}$ of the space.

Part A

The first problem that the center is looking to solve is to determine how much renters should pay. Before they can determine an actual price, the planners must be able to rank the renters in order of the amount of space they are renting. List each renter client from greatest to least with respect to the amount of space that they are renting.

1. Cycles
2. Noodleheads
3. Dora's
4. G-Street
5. Nerd Nation
6. Green Door

Part B

The River Bank planners need to draw up a floor plan of the building to see how each space lines up. So far, the architects have come up with the floor plan shown below that divides the rectangular space into equal-sized parts that will be divided among the clients. Color in the squares to show a way in which the space for each client can be laid out.

Performance Task (continued)

Part C

The clients want to see another variation of the floor plan you made. Draw an additional floor plan and color it in to show each client's space.

Any way that uses the right number of squares will work

Part D

Having gone over the previous plans, the clients now want the architect to come up with as many rectangular plans for the entire complex as possible. How many different rectangles can you draw? Make a sketch of each rectangle showing the individual squares as you drew above. Each rectangle must be at least 3 squares in width.

3

Part E

The builders have decided to charge the renters between $850 and $1,000 per square unit as shown in the diagram in Part B. Determine the least and greatest amount of rent each client will pay.

	Least	Greatest
Green Door Theater	21,250	25,000
G-Street Apps		10,000
Nerd Nation Fitness		14,000
Dora's Cupcakes		6,000
Noodlehead's		3,000
Bike Geek Cycle Shop		

Chapter 8 Performance Task Student Model 4

Performance Task

NAME _____ DATE _____

SCORE _____

The River Bank Market Center

River Bank Market is an urban shopping and entertainment center that is still in its planning stage. River Bank has sold parts of its 1000 square meter space to 6 different client groups. Nerd Nation Fitness has purchased $\frac{7}{30}$ of the space for its center. G-Street Apps has an area equal to $\frac{1}{6}$ of the total space. Bike Geek Cycle Shop has $\frac{1}{30}$ of the center's total space for a shop. Dora's Unbelievably Great Cupcakes has $\frac{1}{10}$ of the space. The Green Door Theater has $\frac{5}{12}$ of the total space. Noodlehead's has $\frac{1}{20}$ of the space.

Part A

The first problem that the center is looking to solve is to determine how much renters should pay. Before they can determine an actual price, the planners must be able to rank the renters in order of the amount of space they are renting. List each renter client from greatest to least with respect to the amount of space that they are renting.

$\frac{5}{12} > \frac{7}{30} > \frac{1}{6} > \frac{1}{10} > \frac{1}{20} > \frac{1}{30}$

Part B

The River Bank planners need to draw up a floor plan of the building to see how each space lines up. So far, the architects have come up with the floor plan shown below that divides the rectangular space into equal-sized parts that will be divided among the clients. Color in the squares to show a way in which the space for each client can be laid out.

All of the squares will be colored. All 60 are needed.

Performance Task (continued)

Part C

The clients want to see another variation of the floor plan you made. Draw an additional floor plan and color it in to show each client's space.

Same as B.

Part D

Having gone over the previous plans, the clients now want the architect to come up with as many rectangular plans for the entire complex as possible. How many different rectangles can you draw? Make a sketch of each rectangle showing the individual squares as you drew above. Each rectangle must be at least 3 squares in width.

3 rectangles

Part E

The builders have decided to charge the renters between $850 and $1,000 per square unit as shown in the diagram in Part B. Determine the least and greatest amount of rent each client will pay.

	Least	Greatest
Green Door Theater		
G-Street Apps		
Nerd Nation Fitness		
Dora's Cupcakes		
Noodlehead's		
Bike Geek Cycle Shop		

$1,000   60,000

Chapter 9 Performance Task Rubric

Page 153 • Ask Omar

Task Scenario
Students will determine whether various measuring devices can be used to measure specific quantities of flour. To find solutions to the problems, students will add, subtract, and multiply fractions and mixed numbers using models and numbers.

Depth of Knowledge	DOK2, DOK3

Part	Maximum Points	Scoring Rubric
A	3	Full Credit: Students create a model to find that Betty needs eight $\frac{2}{3}$ lb measuring cups full of flour to measure the $5\frac{1}{3}$ pounds. Students also need to write a response to Betty explaining how they found their solution. The response should show that adding $\frac{2}{3}$ of pound 8 times will come out to $5\frac{1}{3}$. Partial Credit (2 points) will be given for a correct solution and a correct model, but no response, or an inadequate response to Betty. Partial Credit (2 points) will be given for a correct solution and response to Betty, but no model. Partial Credit (1 point) will be given for a correct solution and no model or response to Betty. No credit will be given for an incorrect solution to the problem.
B	2	Full Credit: Students accurately subtract $2\frac{2}{3}$ from $5\frac{1}{3}$, then correctly find that Gabby needs 16 measuring cups that each hold $\frac{1}{6}$ of a pound. Partial Credit (1 point) will be given for finding the correct number of pounds of flour that Gabby needs but not the number of measuring cups that she needs. No credit will be given for an incorrect solution to the problem.
C	3	Full Credit: Students correctly determine that a $\frac{4}{5}$ lb measuring cup can measure 8 pounds exactly, but not a $\frac{3}{5}$ lb cup. Students also explain how they found their answer, either by multiplying $10 \times \frac{4}{5}$ to get $\frac{40}{5}$, or 8 cups of flour, or by repeating addition of $\frac{4}{5}$ to get 8 cups of flour. Partial Credit (1 point) will be given for correctly determining that Ahmed should use the $\frac{4}{5}$ lb cup but not fully explaining why that cup will work and the $\frac{3}{5}$ lb cup will not work. No credit will be given for an incorrect solution to the problem.

Chapter 9 Performance Task Rubric, continued

Part	Maximum Points	Scoring Rubric
D	2	Full Credit: Students correctly add $5\frac{1}{3}$ four times to obtain a total of $21\frac{1}{3}$ pounds of flour. Students then explain how their repeated addition allowed them to find a recipe for 4 times as many people **OR** Students correctly obtain a total of $21\frac{1}{3}$ pounds of flour using an inventive method, such as multiplying $4 \times 5\frac{1}{3}$ as $(4 \times 5) + (4 \times \frac{1}{3})$ then adding the products. Students then explain how their invented multiplication method allowed them to find a recipe for 4 times as many people as originally planned. Partial Credit (1 point) will be given for a correct solution but no explanation, or an explanation that is incomplete. No credit will be given for an incorrect solution to the problem.
TOTAL	10	

Chapter 9 Performance Task Student Model 1

NAME _____ DATE _____ SCORE _____

Performance Task

Ask Omar

Omar is the author of the website, Ask Omar the Baker, which answers questions about baking. Here is today's question:

Hey Omar, My recipe for biscuits calls for $5\frac{1}{3}$ pounds of flour. But I have only one measuring cup and it measures $\frac{2}{3}$ of a pound of flour at a time. How can I figure out how to measure $5\frac{1}{3}$ pounds of flour using my measuring cup? Signed, Biscuit Betty

Write your answers on another piece of paper. Show all your work to receive full credit.

Part A

Use a model to show how Betty can measure the $5\frac{1}{3}$ pounds of flour using her $\frac{2}{3}$ lb measuring cup. Then write a response from Omar to Betty describing how she can solve the problem.

[student drawing of 8 rectangles, each divided and shaded]

Betty,
Divide whole cups into thirds. Then group two thirds together until you have $5\frac{1}{3}$ pounds. It takes 8 of the $\frac{2}{3}$ pound measures to make $5\frac{1}{3}$ pounds.

Performance Task (continued)

Part B

Gourmet Gabby writes that her recipe uses $2\frac{2}{3}$ fewer pounds of flour than Biscuit Betty's recipe. Gabby's measuring cup holds $\frac{1}{6}$ of a pound. How many measuring cups full of flour does Gabby need?

$5\frac{1}{3} - 2\frac{2}{3} = \frac{16}{3} - \frac{8}{3} = \frac{8}{3} = 2\frac{2}{3}$

$\frac{2\frac{2}{3}}{1+1+\frac{2}{3}}$
$\frac{6}{6} \frac{6}{6} \frac{4}{6}$ 16 measuring cups

Part C

Ahmed writes that he needs to measure 8 pounds of flour. To make an exact measurement with no flour left over, which measuring cup should Ahmed use, one that holds $\frac{3}{5}$ pound or $\frac{4}{5}$ pound? Explain.

$10 \times \frac{4}{5} = \frac{40}{5} = 8$ pounds

$13 \times \frac{3}{5} = \frac{39}{5} = 7\frac{4}{5}$
$14 \times \frac{3}{5} = \frac{42}{5} = 8\frac{2}{5}$

Use the $\frac{4}{5}$ pound measure.

Part D

Toni Mahoney is using Biscuit Betty's recipe that requires $5\frac{1}{3}$ pounds of flour to make biscuits for her brunch. Suddenly, Toni learns that 4 times as many people are coming to her brunch than originally planned. How much flour will Toni need to make her biscuits? Explain how you got your answer.

$5\frac{1}{3} + 5\frac{1}{3} + 5\frac{1}{3} + 5\frac{1}{3} = 21\frac{1}{3}$

Chapter 9 Performance Task Student Model 2

NAME _____ DATE _____

SCORE _____

Performance Task

Ask Omar

Omar is the author of the website, Ask Omar the Baker, which answers questions about baking. Here is today's question:

Hey Omar, My recipe for biscuits calls for $5\frac{1}{3}$ pounds of flour. But I have only one measuring cup and it measures $\frac{2}{3}$ of a pound of flour at a time. How can I figure out how to measure $5\frac{1}{3}$ pounds of flour using my measuring cup? Signed, Biscuit Betty

Write your answers on another piece of paper. Show all your work to receive full credit.

Part A

Use a model to show how Betty can measure the $5\frac{1}{3}$ pounds of flour using her $\frac{2}{3}$ lb measuring cup. Then write a response from Omar to Betty describing how she can solve the problem.

$\frac{2}{3} + \frac{2}{3} + \frac{2}{3} + \frac{2}{3} + \frac{2}{3} + \frac{2}{3} + \frac{2}{3} + \frac{2}{3} = \frac{16}{3} = 5\frac{1}{3}$

Betty,
To find out how many $\frac{2}{3}$ pound measures you need, add $\frac{2}{3}$ together until you get $5\frac{1}{3}$.

Performance Task (continued)

Part B

Gourmet Gabby writes that her recipe uses $2\frac{2}{3}$ fewer pounds of flour than Biscuit Betty's recipe. Gabby's measuring cup holds $\frac{1}{6}$ of a pound. How many measuring cups full of flour does Gabby need?

$5\frac{1}{3} - 2\frac{2}{3} = \frac{4}{3} - 2\frac{2}{3} = 2\frac{2}{3}$

6 + 6 + 4 = 16 measuring cups

Part C

Ahmed writes that he needs to measure 8 pounds of flour. To make an exact measurement with no flour left over, which measuring cup should Ahmed use, one that holds $\frac{3}{5}$ pound or $\frac{4}{5}$ pound? Explain.

$\frac{4}{5} + \frac{4}{5} + \frac{4}{5} + \frac{4}{5} + \frac{4}{5} + \frac{4}{5} + \frac{4}{5} + \frac{4}{5} + \frac{4}{5} + \frac{4}{5} = \frac{40}{5} = 8$

Use the $\frac{4}{5}$ pound measure 10 times.

Part D

Toni Mahoney is using Biscuit Betty's recipe that requires $5\frac{1}{3}$ pounds of flour to make biscuits for her brunch. Suddenly, Toni learns that 4 times as many people are coming to her brunch than originally planned. How much flour will Toni need to make her biscuits? Explain how you got your answer.

$5\frac{1}{3} \times 4 = 5 \times 4 + \frac{1}{3} \times 4 = 20 + \frac{4}{3} = 21\frac{1}{3}$

Chapter 9 Performance Task Student Model 3

NAME _____ **DATE** _____ **SCORE** _____

Performance Task

Ask Omar

Omar is the author of the website, Ask Omar the Baker, which answers questions about baking. Here is today's question:

Hey Omar, My recipe for biscuits calls for $5\frac{1}{3}$ pounds of flour. But I have only one measuring cup and it measures $\frac{2}{3}$ of a pound of flour at a time. How can I figure out how to measure $5\frac{1}{3}$ pounds of flour using my measuring cup? Signed, Biscuit Betty

Write your answers on another piece of paper. Show all your work to receive full credit.

Part A

Use a model to show how Betty can measure the $5\frac{1}{3}$ pounds of flour using her $\frac{2}{3}$ lb measuring cup. Then write a response from Omar to Betty describing how she can solve the problem.

8 groups of $\frac{2}{3}$ make $5\frac{1}{3}$

Performance Task (continued)

Part B

Gourmet Gabby writes that her recipe uses $2\frac{2}{3}$ fewer pounds of flour than Biscuit Betty's recipe. Gabby's measuring cup holds $\frac{1}{6}$ of a pound. How many measuring cups full of flour does Gabby need?

$3\frac{5}{3} - 2\frac{2}{3} = 3\frac{1}{3} - \frac{2}{3} = 2\frac{2}{3}$

$\frac{2}{3} + \frac{2}{3} + \frac{2}{3} = 2\frac{2}{3}$

$\frac{2}{3} + \frac{2}{3} + \frac{2}{3} = 2\frac{2}{3}$

4 measuring cups

Part C

Ahmed writes that he needs to measure 8 pounds of flour. To make an exact measurement with no flour left over, which measuring cup should Ahmed use, one that holds $\frac{3}{5}$ pound or $\frac{4}{5}$ pound? Explain.

$\frac{3}{5} + \frac{3}{5} + \frac{3}{5} + \frac{3}{5} + \frac{3}{5} = 3$

$\frac{3}{5} + \frac{3}{5} + \frac{3}{5} + \frac{3}{5} + \frac{3}{5} = 3$

5 groups = 3
10 groups = 6
Can't get 8
Use the $\frac{4}{5}$ pound cup.

Part D

Toni Mahoney is using Biscuit Betty's recipe that requires $5\frac{1}{3}$ pounds of flour to make biscuits for her brunch. Suddenly, Toni learns that 4 times as many people are coming to her brunch than originally planned. How much flour will Toni need to make her biscuits? Explain how you got your answer.

More than 20 pounds

Chapter 9 Performance Task Student Model 4

NAME _____ DATE _____

SCORE _____

Performance Task

Ask Omar

Omar is the author of the website, Ask Omar the Baker, which answers questions about baking. Here is today's question:

Hey Omar, My recipe for biscuits calls for $5\frac{1}{3}$ pounds of flour. But I have only one measuring cup and it measures $\frac{2}{3}$ of a pound of flour at a time. How can I figure out how to measure $5\frac{1}{3}$ pounds of flour using my measuring cup? Signed, Biscuit Betty

Write your answers on another piece of paper. Show all your work to receive full credit.

Part A

Use a model to show how Betty can measure the $5\frac{1}{3}$ pounds of flour using her $\frac{2}{3}$ lb measuring cup. Then write a response from Omar to Betty describing how she can solve the problem.

Add $\frac{2}{3}$ until you get $5\frac{1}{3}$.

Performance Task (continued)

Part B

Gourmet Gabby writes that her recipe uses $2\frac{2}{3}$ fewer pounds of flour than Biscuit Betty's recipe. Gabby's measuring cup holds $\frac{1}{6}$ of a pound. How many measuring cups full of flour does Gabby need?

1 pound = 6 measures
2 pounds = 12 measures
$\frac{2}{3}$ pound = 4 measures

Part C

Ahmed writes that he needs to measure 8 pounds of flour. To make an exact measurement with no flour left over, which measuring cup should Ahmed use, one that holds $\frac{3}{5}$ pound or $\frac{4}{5}$ pound? Explain.

Use the $\frac{3}{5}$ pound measuring cup.

Part D

Toni Mahoney is using Biscuit Betty's recipe that requires $5\frac{1}{3}$ pounds of flour to make biscuits for her brunch. Suddenly, Toni learns that 4 times as many people are coming to her brunch than originally planned. How much flour will Toni need to make her biscuits? Explain how you got your answer.

Multiply $5\frac{1}{3} \times 4$.

Chapter 10 Performance Task Rubric

Page 155 • Wendy's Weather Station

Task Scenario

Students will use a weather map to record and interpret data in table form. To find solutions to problems students must convert back and forth from fraction to decimal form for a variety of numbers. They must also use reasoning skills to solve problems that involve addition, identifying equivalencies, and comparing and ranking numbers.

Depth of Knowledge	DOK2, DOK3				
Part	**Maximum Points**	**Scoring Rubric**			
A	2	Full Credit: Students accurately record each rain total from the map onto the table and correctly write its equivalent decimal or fractional quantity.				

Location	Rio Rancho	North Valley	Old Town	Downtown	Southtown
Rain (in) fraction	$\frac{60}{100}$	$\frac{6}{10}$	$\frac{9}{100}$	$\frac{9}{100}$	$\frac{6}{10}$
Rain (in) decimal	0.60	0.6	0.09	0.09	0.6
Location	West Side	South Valley	Petroglyph	Sandia	Four Hills
Rain (in) fraction	$\frac{63}{100}$	$\frac{72}{100}$	$\frac{6}{100}$	$\frac{49}{100}$	$\frac{7}{10}$
Rain (in) decimal	0.63	0.72	0.06	0.49	0.7

Partial Credit (1 point) will be given for recording rain totals from the map accurately, but making one error in writing equivalent decimal or fractional numbers.

No credit will be given for recording totals incorrectly or making more than one error in the table.

Grade 4 • Chapter 10 Performance Task Rubric

Chapter 10 Performance Task Rubric, continued

Part	Maximum Points	Scoring Rubric
B	2	Full Credit: Students accurately rank the totals from greatest to least and mark ties correctly. | Location	Rio Rancho	North Valley	Old Town	Downtown	Southtown
---	---	---	---	---	---
Rank	4 (tie)	4 (tie)	6 (tie)	6 (tie)	4 (tie)
Location	West Side	South Valley	Petroglyph	Sandia	Four Hills
Rank	3	1	7	5	2	 Partial Credit (1 point) will be given for accurately ranking the rain totals but listing them from least to greatest rather than from greatest to least. Partial Credit (1 point) will be given for accurately ranking rain totals but failing to identify ties in which more than one location had the same rain total. No credit will be given for incorrect ranking of the totals from the map.
C	3	Full Credit: Students add the correct Thursday rain totals and correctly find the sums for both days. | Location	Rio Rancho	North Valley	Old Town	Downtown	Southtown
---	---	---	---	---	---
Rain (in) Wednesday	0.60	0.6	0.09	0.09	0.6
Rain (in) Thursday	0.15	0.18	0.18	0.18	0.18
Total (in) Weds/Thurs	0.75	0.78	0.27	0.27	0.78
Location	West Side	South Valley	Petroglyph	Sandia	Four Hills
Rain (in) Wednesday	0.63	0.72	0.06	0.49	0.7
Rain (in) Thursday	0.15	0.15	0.15	0.18	0.18
Total (in) Weds/Thurs	0.78	0.87	0.21	0.67	0.88	 Partial Credit (2 points) for correctly filling out the table but expressing totals in fractional rather than decimal form. Partial Credit (1 point) will be given for recording rain totals from the map and committing a single error in recording which causes a sum to be inaccurate. No credit will be given for committing more than one error for the sums.
TOTAL	7	

Chapter 10 Performance Task Student Model 1

NAME _____ DATE _____ SCORE _____

Performance Task

Wendy's Weather Station

Wendy's Weather Website collects data from weather volunteers. The map shows rain totals in inches for Wednesday. Wendy would like to post the totals, but she sees that some volunteers gave their total in fraction form and some gave totals in decimal form.

Map shows:
- Rio Rancho 60/100 in
- North Valley 0.6 in
- Sandia 0.49 in
- Four Hills 7/10 in
- West Side 63/100 in
- Old Town 0.09 in
- Downtown 9/100 in
- Southtown 6/10 in
- Petroglyph 0.06 in
- South Valley 0.72 in
- Rio Grande River

Write your answers on another piece of paper. Show all your work to receive full credit.

Part A
Fill in the table to post rain totals for Wednesday.

Location	Rio Rancho	North Valley	Old Town	Downtown	Southtown
Rain (in) fraction	60/100	6/10	9/100	9/100	6/10
Rain (in) decimal	0.60	0.6	0.09	0.09	0.6

Location	West Side	South Valley	Petroglyph	Sandia	Four Hills
Rain (in) fraction	63/100	72/100	6/100	49/100	7/10
Rain (in) decimal	0.63	0.72	0.06	0.49	0.7

Performance Task (continued)

Part B
Rank the rain totals from greatest to least in the table. Indicate locations that are tied and have the same amount of rainfall.

Location	Rio Rancho	North Valley	Old Town	Downtown	Southtown
Rank	4 (tie)	4 (tie)	6 (tie)	6 (tie)	4 (tie)

Location	West Side	South Valley	Petroglyph	Sandia	Four Hills
Rank	3	1	7	5	2

Part C
On Thursday it rained 0.15 inches at every location west of the river and 18/100 inches at all locations east of the river. Find decimal totals for each location for Wednesday and Thursday combined.

Location	Rio Rancho	North Valley	Old Town	Downtown	Southtown
Rain (in) Wednesday	0.60	0.6	0.09	0.09	0.6
Rain (in) Thursday	0.15	0.18	0.18	0.18	0.18
Total (in) Weds/Thurs	0.75	0.78	0.27	0.27	0.78

Location	West Side	South Valley	Petroglyph	Sandia	Four Hills
Rain (in) Wednesday	0.63	0.72	0.06	0.49	0.7
Rain (in) Thursday	0.15	0.15	0.15	0.18	0.18
Total (in) Weds/Thurs	0.78	0.87	0.21	0.67	0.88

Chapter 10 Performance Task Student Model 2

NAME _____ DATE _____

SCORE _____

Performance Task

Wendy's Weather Station

Wendy's Weather Website collects data from weather volunteers. The map shows rain totals in inches for Wednesday. Wendy would like to post the totals, but she sees that some volunteers gave their total in fraction form and some gave totals in decimal form.

Map locations:
- Rio Rancho 60/100 in
- North Valley 0.6 in
- West Side 63/100 in
- Old Town 0.09 in
- Sandia 0.49 in
- Downtown 9/100 in
- Petroglyph 0.06 in
- Southtown 6/10 in
- South Valley 0.72 in
- Four Hills 7/10 in
- Rio Grande River

Write your answers on another piece of paper. Show all your work to receive full credit.

Part A

Fill in the table to post rain totals for Wednesday.

Location	Rio Rancho	North Valley	Old Town	Downtown	Southtown
Rain (in) fraction	60/100	6/10	9/100	9/100	6/10
Rain (in) decimal	0.6	0.6	0.09	0.09	0.6

Location	West Side	South Valley	Petroglyph	Sandia	Four Hills
Rain (in) fraction	63/100	72/100	6/100	49/100	7/10
Rain (in) decimal	0.63	0.72	0.06	0.49	0.7

Performance Task (continued)

Part B

Rank the rain totals from greatest to least in the table. Indicate locations that are tied and have the same amount of rainfall.

Location	Rio Rancho	North Valley	Old Town	Downtown	Southtown
Rank	4	4	6	6	4

Location	West Side	South Valley	Petroglyph	Sandia	Four Hills
Rank	3	1	7	5	2

Part C

On Thursday it rained 0.15 inches at every location west of the river and 0.18 inches at all locations east of the river. Find decimal totals for each location for Wednesday and Thursday combined.

Location	Rio Rancho	North Valley	Old Town	Downtown	Southtown
Rain (in) Wednesday	0.6	0.6	0.09	0.09	0.6
Rain (in) Thursday	0.15	0.18	0.18	0.18	0.18
Total (in) Weds/Thurs	0.21	0.24	0.27	0.27	0.24

Location	West Side	South Valley	Petroglyph	Sandia	Four Hills
Rain (in) Wednesday	0.63	0.72	0.06	0.49	0.7
Rain (in) Thursday	0.15	0.15	0.15	0.18	0.18
Total (in) Weds/Thurs	0.88	0.87	0.21	0.67	0.78

Chapter 10 Performance Task Student Model 3

Performance Task

Wendy's Weather Station

Wendy's Weather Website collects data from weather volunteers. The map shows rain totals in inches for Wednesday. Wendy would like to post the totals, but she sees that some volunteers gave their total in fraction form and some gave their totals in decimal form.

Map locations:
- Rio Rancho 60/100 in
- North Valley 0.6 in
- Sandia 0.49 in
- Four Hills 7/10 in
- West Side 63/100 in
- Old Town 0.09 in
- Downtown 9/100 in
- Southtown 6/10 in
- Petroglyph 0.06 in
- South Valley 0.72 in
- Rio Grande River

Write your answers on another piece of paper. Show all your work to receive full credit.

Part A

Fill in the table to post rain totals for Wednesday.

Location	Rio Rancho	North Valley	Old Town	Downtown	Southtown
Rain (in) fraction	60/100	6/10	9/100	9/100	6/10
Rain (in) decimal	0.60	0.6	0.09	0.09	0.6

Location	West Side	South Valley	Petroglyph	Sandia	Four Hills
Rain (in) fraction	63/100	72/100	6/100	49/100	7/10
Rain (in) decimal	0.63	0.72	0.06	0.49	0.7

Performance Task (continued)

Part B

Rank the rain totals from greatest to least in the table. Indicate locations that are tied and have the same amount of rainfall.

Location	Rio Rancho	North Valley	Old Town	Downtown	Southtown
Rank	4(tie)	4(tie)	2(tie)	2(tie)	4(tie)

Location	West Side	South Valley	Petroglyph	Sandia	Four Hills
Rank	5	1	7	3	6

Part C

On Thursday it rained 0.15 inches at every location west of the river and 18/100 inches at all locations east of the river. Find decimal totals for each location for Wednesday and Thursday combined.

Location	Rio Rancho	North Valley	Old Town	Downtown	Southtown
Rain (in) Wednesday	0.60	6/10	9/100	9/100	6/10
Rain (in) Thursday	0.15	18/100	18/100	18/100	18/100
Total (in) Weds/Thurs	0.75	24/100	27/100	27/100	24/100

Location	West Side	South Valley	Petroglyph	Sandia	Four Hills
Rain (in) Wednesday	0.63	0.72	0.06	49/100	7/10
Rain (in) Thursday	0.15	0.15	0.15	18/100	18/100
Total (in) Weds/Thurs	0.78	0.87	0.21	67/100	25/100

Chapter 10 Performance Task Student Model 4

NAME _____ DATE _____ SCORE _____

Performance Task

Wendy's Weather Station

Wendy's Weather Website collects data from weather volunteers. The map shows rain totals in inches for Wednesday. Wendy would like to post the totals, but she sees that some volunteers gave their total in fraction form and some gave totals in decimal form.

Map:
- Rio Rancho 60/100 in
- North Valley 0.6 in
- Sandia 0.49 in
- Four Hills 7/10 in
- West Side 63/100 in
- Old Town 0.09 in
- Downtown 9/100 in
- Southtown 6/10 in
- Petroglyph 0.06 in
- South Valley 0.72 in
- Rio Grande River

Write your answers on another piece of paper. Show all your work to receive full credit.

Part A
Fill in the table to post rain totals for Wednesday.

Location	Rio Rancho	North Valley	Old Town	Downtown	Southtown
Rain (in) fraction	60/100			9/100	6/10
Rain (in) decimal		0.6	0.09		

Location	West Side	South Valley	Petroglyph	Sandia	Four Hills
Rain (in) fraction	63/100				7/10
Rain (in) decimal		0.72	0.06	0.49	

Part B
Rank the rain totals from greatest to least in the table. Indicate locations that are tied and have the same amount of rainfall.

Location	Rio Rancho	North Valley	Old Town	Downtown	Southtown
Rank	4	7	6	6	4

Location	West Side	South Valley	Petroglyph	Sandia	Four Hills
Rank	2	1	7	5	3

Part C
On Thursday it rained 0.15 inches at every location west of the river and 18/100 inches at all locations east of the river. Find decimal totals for each location for Wednesday and Thursday combined.

Location	Rio Rancho	North Valley	Old Town	Downtown	Southtown
Rain (in) Wednesday	60/100	0.6	0.09	9/100	6/10
Rain (in) Thursday	0.15	18/100	18/100	18/100	18/100
Total (in) Weds/Thurs				27/100	78/100

Location	West Side	South Valley	Petroglyph	Sandia	Four Hills
Rain (in) Wednesday	63/100	0.72	0.06	0.49	7/10
Rain (in) Thursday	0.15	0.15	0.15	18/100	18/100
Total (in) Weds/Thurs		0.87	0.21		88/100

Chapter 11 Performance Task Rubric

Page 157 • Manny the Mover

Task Scenario
Students create a floor plan model and use tile models to determine the number of boxes that would fit in a floor plan. To create the dimensions of the floor plan, students must convert customary measurements from the units given to a common unit for all measurements. Students must use their model to draw conclusions and use spatial reasoning to solve problems.

Depth of Knowledge	DOK2, DOK3	
Part	**Maximum Points**	**Scoring Rubric**
A	2	Full Credit: Students create a useful floor plan drawing that is correctly labeled and includes a grid. Partial Credit (1 point) will be given for drawing a model that does not include a grid but is accurately scaled. No credit will be given for a drawing that has incorrect dimensions.
B	3	Full Credit: Students accurately show three models of how to arrange boxes on the floor plan. Note that there are a variety of different ways to do this using boxes that are laid out on their 2 foot × 3 foot sides. All of these arrangements will show the same number of boxes, 32. If laying the boxes on their 2 foot by 2 foot side, the layer contains 48 boxes. Partial Credit (2 points) will be given for providing 2 out of 3 correct diagrams. Partial Credit (1 point) will be given for providing one correct diagram. No credit will be given for providing no correct diagrams.
C	3	Full Credit: Students find the correct number of boxes for Plan 1 (32), Plan 2 (32), and Plan 3 (48). Partial Credit (1 point) will be given for finding the correct number of boxes for 2 of the 3 plans. No credit will be given for finding the correct number of boxes for fewer than 2 of the 3 plans.

Chapter 11 Performance Task Rubric, continued

Part	Maximum Points	Scoring Rubric
D	3	Full Credit: Students find the correct number of boxes for Plan 1 (96), Plan 2 (96), or Plan 3 (96). Partial Credit (1 point) will be given for finding the correct number of boxes for 2 of the 3 plans. No credit will be given for finding the correct number of boxes for fewer than 2 of the 3 plans.
E	4	Full Credit: Students correctly recognize that the 108-inch ceiling (9 feet) allows 3 layers of boxes set on their ends to be stacked, giving a total of 144 boxes, while the boxes on their sides will stack in 4 layers and give a total of 128 boxes. Students should also recognize that stacking boxes on their sides will leave a 1-foot gap. Some students may discover that they can combine Plan 3 with one of the other plans by first packing 3 layers of boxes on their sides, giving 3 × 32 or 96 boxes, then packing 1 layer of boxes on their ends, adding 48 boxes, to give a total of 144 boxes with no gap. Partial Credit (1 point) will be given for correctly determining the total number of boxes for 2 of the 3 plans. No credit will be given for determining the correct number of boxes for fewer than 2 of the 3 plans.
TOTAL	15	

Chapter 11 Performance Task Student Model 1

Performance Task

Manny the Mover

On the side of Manny the Mover's truck, it says he can move "anything, anywhere, anytime." Manny's current order features boxes that measure 3 feet by 2 feet by 2 feet. The floor of Manny's truck bed measures 8 yards in length and 96 inches in width.

Write your answers on another piece of paper. Show all your work to receive full credit.

Part A

Draw and label a floor plan model of Manny's truck bed. Model tiles to represent boxes are shown. Use a grid pattern to show each square unit in your diagram.

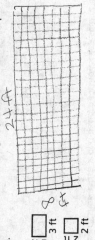

Part B

Show 3 different plans labeled Plan 1, Plan 2, and Plan 3 for packing 1 layer of boxes on the truck. Assume that you can squeeze boxes into tight spaces. For example, you can squeeze a 2-foot wide box into a 2-foot space.

Plan 1
2×3

Plan 3
2×2

Plan 2
3×2

Performance Task (continued)

Part C

Which plan from Part B allows you to pack the greatest number of boxes? Explain.

The plan with the 2×2 face down fits 48 boxes, the others fit 32 boxes.

Part D

The height of Manny's cargo space is 2 yards and 12 inches. If Manny wants to choose the plan that can fit the most boxes with the smallest amount of gaps, which plan should choose? Stack the boxes in layers of equal height. How many boxes will each plan allow? Will there be any gaps? Explain.

2 yards 12 inches = 7 feet
The 32 box plans fit 3 layers for 96 boxes.
The 48 box plan fits 2 layers for 96 boxes.
All are equal, all have a foot gap.

Part E

Manny has just discovered that his special tall truck with a 108-inch ceiling is available. If he uses the tall truck, does this change the number of boxes that he can pack? Explain.

108 inches = 9 feet
32 box plans - 4 layers (1 foot gap) = 128 boxes
48 box plan - 3 layers (no gap) = 144 boxes

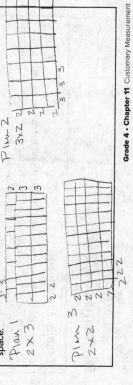

Chapter 11 Performance Task Student Model 2

NAME _____ DATE _____

SCORE _____

Performance Task

Manny the Mover

On the side of Manny the Mover's truck, it says he can move "anything, anywhere, anytime." Manny's current order features boxes that measure 3 feet by 2 feet by 2 feet. The floor of Manny's truck bed measures 8 yards in length and 96 inches in width.

Write your answers on another piece of paper. Show all your work to receive full credit.

Part A

Draw and label a floor plan model of Manny's truck bed. Model tiles to represent boxes are shown. Use a grid pattern to show each square unit in your diagram.

8 yards = 24 feet
96 inches = 8 feet

[grid diagram labeled 24 and 8]

3 ft / 2 ft (dark)
2 ft / 2 ft (light)

Part B

Show 3 different plans labeled Plan 1, Plan 2, and Plan 3 for packing 1 layer of boxes on the truck. Assume that you can squeeze boxes into tight spaces. For example, you can squeeze a 2-foot wide box into a 2-foot space.

4 by 12 = 48 boxes
8 by 4 = 32 boxes

Part C

Which plan from Part B allows you to pack the greatest number of boxes? Explain.

The one with the boxes 2 and 3 ft high.

Part D

The height of Manny's cargo space is 2 yards and 12 inches. If Manny wants to choose the plan that can fit the most boxes with the smallest amount of gaps, which plan should choose? Stack the boxes in layers of equal height. How many boxes will each plan allow? Will there be any gaps? Explain.

7 feet

6 ft tall, 2 layers → 96 boxes
6 ft tall, 3 layers → 96 boxes

Part E

Manny has just discovered that his special tall truck with a 108-inch ceiling is available. If he uses the tall truck, does this change the number of boxes that he can pack? Explain.

108 in = 9 ft
Now he can fit 3 layers of the 48 box plan with no gaps.
3 × 48 = 144 boxes

Chapter 11 Performance Task Student Model 3

NAME _____ DATE _____

SCORE _____

Performance Task

Manny the Mover

On the side of Manny the Mover's truck, it says he can move "anything, anywhere, anytime." Manny's current order features boxes that measure 3 feet by 2 feet by 2 feet. The floor of Manny's truck bed measures 8 yards in length and 96 inches in width.

Write your answers on another piece of paper. Show all your work to receive full credit.

Part A

Draw and label a floor plan model of Manny's truck bed. Model tiles to represent boxes are shown. Use a grid pattern to show each square unit in your diagram.

8 yards = 24 feet

24
× 8
192

96 inches = 8 feet

Box 1: 2 ft × 3 ft = 6
Box 2: 2 ft × 2 ft = 4

32 of Box 1
48 of Box 2

Part B

Show 3 different plans labeled Plan 1, Plan 2, and Plan 3 for packing 1 layer of boxes on the truck. Assume that you can squeeze boxes into tight spaces. For example, you can squeeze a 2-foot wide box into a 2-foot space.

33
6)192
−18
   12
  −12
    0

48
4)192
−16
   32
  −32
    0

Performance Task (continued)

Part C

Which plan from Part B allows you to pack the greatest number of boxes? Explain.

Box 2

Part D

The height of Manny's cargo space is 2 yards and 12 inches. If Manny wants to choose the plan that can fit the most boxes with the smallest amount of gaps, which plan should choose? Stack the boxes in layers of equal height. How many boxes will each plan allow? Will there be any gaps? Explain.

6 ft + 1 ft = 7 ft

Box 2: 3 ft + 3 ft = 1 ft gap
48 + 48 = 96 boxes

Box 1: 2 ft + 2 ft + 2 ft + 1 ft gap
3 + 32 + 32 = 96 boxes

Part E

Manny has just discovered that his special tall truck with a 108-inch ceiling is available. If he uses the tall truck, does this change the number of boxes that he can pack? Explain.

Box 2 = 144
Box 1 = 128

Another layer of

Chapter 11 Performance Task Student Model 4

NAME

DATE

SCORE

Performance Task

Manny the Mover

On the side of Manny the Mover's truck, it says he can move "anything, anywhere, anytime." Manny's current order features boxes that measure 3 feet by 2 feet by 2 feet. The floor of Manny's truck bed measures 8 yards in length and 96 inches in width.

Write your answers on another piece of paper. Show all your work to receive full credit.

Part A

Draw and label a floor plan model of Manny's truck bed. Model tiles to represent boxes are shown. Use a grid pattern to show each square unit in your diagram.

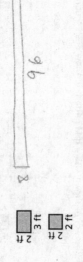

3 ft
2 ft

2 ft
2 ft

8
96

Part B

Show 3 different plans labeled Plan 1, Plan 2, and Plan 3 for packing 1 layer of boxes on the truck. Assume that you can squeeze boxes into tight spaces. For example, you can squeeze a 2-foot wide box into a 2-foot space.

Plan 1: You can fit 4 rows of 3s of the first kind of box.

Plan 2: You can fit 4 rows of 48 of the second kind of box.

Performance Task (continued)

Part C

Which plan from Part B allows you to pack the greatest number of boxes? Explain.

Plan 2 Ans 4×48= 192 boxes.

Part D

The height of Manny's cargo space is 2 yards and 12 inches. If Manny wants to choose the plan that can fit the most boxes with the smallest amount of gaps, which plan should choose? Stack the boxes in layers of equal height. How many boxes will each plan allow? Will there be any gaps? Explain.

3 layers of plan 1 (384)
or
2 layers of plan 2 (384)

Part E

Manny has just discovered that his special tall truck with a 108-inch ceiling is available. If he uses the tall truck, does this change the number of boxes that he can pack? Explain.

Yes, the truck is taller, so more boxes will fit.

Page 159 • Frog House

Task Scenario Students create a floor plan model of the Frog House using metric units. Students then use their model to place individual figures on the floor plan that meet a variety of spatial characteristics. Students must draw conclusions and use mathematical and spatial reasoning to solve problems.		
Depth of Knowledge		DOK2, DOK3
Part	**Maximum Points**	**Scoring Rubric**
A	3	Full Credit: Students find the correct dimensions for each habitat in meters. This will show the following dimensions: bullfrogs: 4 m × 6 m, leopard frogs: 4 m × 5 m, toads: 1.5 m × 2.5 m, spring peepers: 1 m × 4 m, poison dart frogs: 4 m × 4 m. Partial Credit (2 points) will be given for finding the correct dimensions but dimensions are not given in units of meters. Partial Credit (1 point) will be given for finding the correct dimensions for all but one of the habitats. No credit will be given for more than one set of incorrect dimensions.
B	2	Full Credit: Students create a useful floor plan drawing of the Frog House that is correctly labeled in meters and includes a grid. 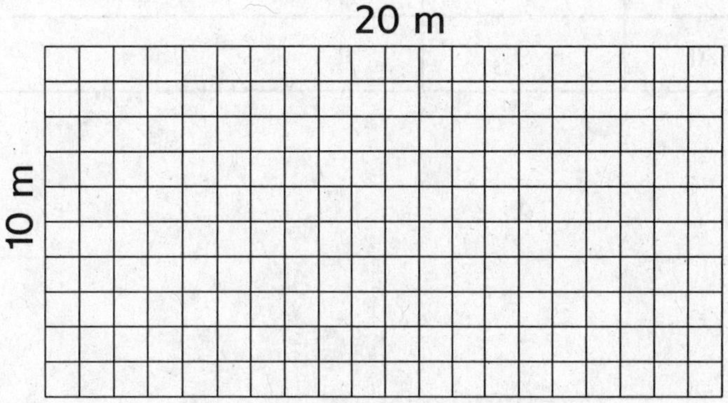 Partial Credit (1 point) will be given for drawing of a model that does not include a grid but is accurately scaled. Partial Credit (1 point) will be given for drawing of a model that is not labeled in meters but its dimensions are correct. No credit will be given for a drawing that has incorrect dimensions.

Chapter 12 Performance Task Rubric, continued

Part	Maximum Points	Scoring Rubric
C	4	Full Credit: Students show a correct arrangement of habitats that have accurate dimensions. All habitats must have a 2 meter distance between one another and a 1 meter distance from the building's outer wall. In addition, the bullfrogs must be 3 meters from an outer wall, the poison dart frogs must be 10 meters from the spring peepers, and the toads must be 10 meters from the spring leopard frogs. *[Diagram: A 20 m by 10 m grid showing habitat arrangement with Poison Dart Frogs (upper left), Spring Peepers (upper right), Bullfrogs (center), Leopard Frogs (right), and Toads (lower left).]* Partial Credit (2 points) will be given for providing a correct diagram that is given in some unit other than meters. Partial Credit (1 point) will be given for making a maximum of 2 minor mistakes within the diagram. No credit will be given for an incorrect diagram that contains more than 2 minor mistakes.
TOTAL	9	

Chapter 12 Performance Task Student Model 1

NAME _____ DATE _____ SCORE _____

Performance Task

Frog House

A zoo is designing a new Frog House that will house leopard frogs, bullfrogs, spring peepers, toads, and poison dart frogs.

Write your answers on another piece of paper. Show all your work to receive full credit.

Part A

Here are the dimensions for each frog's rectangular habitat.
- Bullfrogs: 400 cm by 6000 mm   4m by 6m
- Leopard frogs: Same width as bullfrog, but 1 meter shorter in length 4m by 5m
- Toads: 250 cm shorter in both length and width than leopard frogs 1.5m by 2.5m
- Spring peepers: 1,500 mm greater in length than toads and 50 cm less in width than toads   1m by 4m
- Poison dart frogs: Same length as spring peepers, but 3 m greater in width than peepers   4m by 4m

Find the dimensions of each frog's habitat in meters.

Performance Task (continued)

Part B

Draw and label a rectangular floor plan for the Frog House that measures 0.02 kilometers by 10 meters. Label your floor plan in meters.

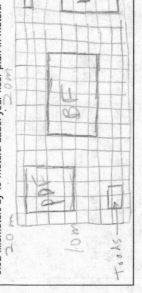

Part C

Here are the special requirements for the habitats. Draw in each habitat on your floor plan.
- All habitats: Must have at least a 2 m distance from any other habitat
- All habitats: Must have at least a 100 cm distance from any outer wall of building
- Bullfrogs: Must be 3 m from any outer wall of the building
- Poison dart frogs: Must have at least a 1,000 cm distance from spring peepers
- Toads: Must have at least a 1,000 cm distance from leopard frogs

Chapter 12 Performance Task Student Model 2

NAME _____ DATE _____ SCORE _____

Performance Task

Frog House

A zoo is designing a new Frog House that will house leopard frogs, bullfrogs, spring peepers, toads, and poison dart frogs.

Write your answers on another piece of paper. Show all your work to receive full credit.

Part A

Here are the dimensions for each frog's rectangular habitat.
- Bullfrogs: 400 cm by 6000 mm
- Leopard frogs: Same width as bullfrog, but 1 meter shorter in length
- Toads: 250 cm shorter in both length and width than leopard frogs
- Spring peepers: 1,500 mm greater in length than toads and 50 cm less in width than toads
- Poison dart frogs: Same length as spring peepers, but 3 m greater in width than peepers

Find the dimensions of each frog's habitat in meters.

4m × 6m
1.5m × 2.5m
4m × 4m
4m × 5m
1m × 4m

Performance Task (continued)

Part B

Draw and label a rectangular floor plan for the Frog House that measures 0.02 kilometers by 10 meters. Label your floor plan in meters.

[Floor plan drawing: 20 m by 10 m rectangle with habitats labeled PDF, B, LF, SP, toads]

Part C

Here are the special requirements for the habitats. Draw in each habitat on your floor plan.
- All habitats: Must have at least a 2 m distance from any other habitat
- All habitats: Must have at least a 100 cm distance from any outer wall of building
- Bullfrogs: Must be 3 m from any outer wall of the building
- Poison dart frogs: Must have at least a 1,000 cm distance from spring peepers
- Toads: Must have at least a 1,000 cm distance from leopard frogs

Chapter 12 Performance Task Student Model 3

NAME _____ DATE _____

SCORE _____

Performance Task

Frog House

A zoo is designing a new Frog House that will house leopard frogs, bullfrogs, spring peepers, toads, and poison dart frogs.

Write your answers on another piece of paper. Show all your work to receive full credit.

Part A

Here are the dimensions for each frog's rectangular habitat

- Bullfrogs: 400 cm by 6000 mm
- Leopard frogs: Same width as bullfrog, but 1 meter shorter in length
- Toads: 250 cm shorter in both length and width than leopard frogs
- Spring peepers: 1,500 mm greater in length than toads and 50 cm less in width than toads
- Poison dart frogs: Same length as spring peepers, but 3 m greater in width than peepers

Find the dimensions of each frog's habitat in meters.

bullfrogs ✓   leopard ✓   toads ✓
4m by 6m   4m by 5m   1.5m by 2.5m

peepers ✓   dart ✓
3m × 4m   6m × 4m

Performance Task (continued)

Part B

Draw and label a rectangular floor plan for the Frog House that measures 0.02 kilometers by 10 meters. Label your floor plan in meters.

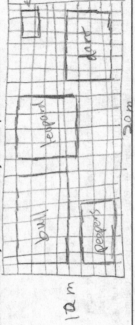

Part C

Here are the special requirements for the habitats. Draw in each habitat on your floor plan.

- All habitats: Must have at least a 2 m distance from any other habitat
- All habitats: Must have at least a 100 cm distance from any outer wall (1m) of building
- Bullfrogs: Must be 3 m from any outer wall of the building
- Poison dart frogs: Must have at least a 1,000 cm distance from spring peepers
- Toads: Must have at least a 1,000 cm distance from leopard frogs

Chapter 12 Performance Task Student Model 4

Performance Task

Frog House

A zoo is designing a new Frog House that will house leopard frogs, bullfrogs, spring peepers, toads, and poison dart frogs.

Write your answers on another piece of paper. Show all your work to receive full credit.

Part A

Here are the dimensions for each frog's rectangular habitat.
- Bullfrogs: 400 cm by 6000 mm
- Leopard frogs: Same width as bullfrog, but 1 meter shorter in length
- Toads: 250 cm shorter in both length and width than leopard frogs
- Spring peepers: 1,500 mm greater in length than toads and 50 cm less in width than toads
- Poison dart frogs: Same length as spring peepers, but 3 m greater in width than peepers

Find the dimensions of each frog's habitat in meters.

BF
4m × 6m

LF
4m × 5m

Toads
3.75m × 4.75m

SP
3.25m × 6.25m

PDF
6.25m × 6.25m

Part B

Draw and label a rectangular floor plan for the Frog House that measures 0.02 kilometers by 10 meters. Label your floor plan in meters.

2m

10m

2m

Part C

Here are the special requirements for the habitats. Draw in each habitat on your floor plan.
- All habitats: Must have at least a 2 m distance from any other habitat
- All habitats: Must have at least a 100 cm distance from any outer wall of building
- Bullfrogs: Must be 3 m from any outer wall of the building
- Poison dart frogs: Must have at least a 1,000 cm distance from spring peepers
- Toads: Must have at least a 1,000 cm distance from leopard frogs

The Frog House is too small.

Chapter 13 Performance Task Rubric

Page 161 • Square Deal

Task Scenario Students make models to show a pattern of squares that increase in size. Students use their models to fill in tables to show how the pattern progresses. Then students analyze their table data and use their data to draw conclusions.		
Depth of Knowledge	DOK2, DOK3	
Part	**Maximum Points**	**Scoring Rubric**
A	2	Full Credit: Students create models that show a progression of squares that increase by one unit on each side. Partial Credit (1 point) will be given for providing at least 4 of the models in the progression. No credit will be given for incorrect models.
B	3	Full Credit: Students fill in the table correctly, describe the pattern generated for both perimeter and area accurately, and determine that area increases at a much faster rate than perimeter. Partial Credit (2 points) will be given for filling in the table correctly, describing the pattern for both perimeter and area accurately, but failing to determine that area increases at a much faster rate than perimeter **OR** for filling in the table correctly, determining that area increases at a much faster rate than perimeter, but failing to describe one of the patterns generated accurately. Partial Credit (1 point) will be given for filling in the table correctly and completing only one of the three remaining tasks correctly. No credit will be given for filling in the table incorrectly **OR** for completing none of the three remaining tasks correctly.

Chapter 13 Performance Task Rubric, continued

Part	Maximum Points	Scoring Rubric
C	3	Full Credit: Students correctly analyze numerical patterns for both perimeter and area. They also accurately describe these patterns in words, recognizing that perimeter increases by a constant amount, while the area increases by more at each stage, changing by an increase of 3, 5, and so on. Partial Credit (2 points) will be given for correctly analyzing numerical patterns for both perimeter and area and accurately describing one of the two, but not both patterns in words. Partial Credit (1 point) will be given for correctly analyzing numerical patterns for both perimeter and area but failing to correctly describe either pattern in words. No credit will be given for failing to analyze the numerical patterns correctly.
D	3	Full Credit: Students correctly complete the table for both perimeter and area. They also accurately describe the patterns in words, recognizing that perimeter doubles between each increment, while area increases by a factor of four. Partial Credit (1 point) will be given for correctly completing the table for both perimeter and area but failing to accurately describe the pattern for either or both quantities. No credit will be given for filling in the table incorrectly.
TOTAL	11	

Chapter 13 Performance Task Student Model 1

Performance Task

Square Deal

Perimeter and area are two ways to measure the size of a figure. In this activity, you will explore patterns of perimeter and area.

Write your answers on another piece of paper. Show all your work to receive full credit.

Part A

Explore a pattern of squares. Create models to show how squares increase in perimeter and area as their side length increases. The first three are done for you. Draw the next seven.

Part B

Now show the numerical patterns of the series. Refer to the models you drew in Part A to fill in the table. In words, describe the pattern for both perimeter and area. Which pattern increases in value faster, perimeter or area? Explain.

Length (ft)	Width (ft)	Perimeter, P (ft)	Area, A (sq ft)
1	1	4	1
2	2	8	4
3	3	12	9
4	4	16	16
5	5	20	25
6	6	24	36
7	7	28	49
8	8	32	64
9	9	36	81
10	10	40	100

Area increases faster the perimeter increases at a regular rate.

Performance Task (continued)

Part C

Now, analyze the patterns for perimeter and area. Find the differences in perimeter and area as the squares increase in side length. Fill in the table. Then describe how each pattern changes.

Length (ft)	Width (ft)	Perimeter, P (ft)	Perimeter difference	Area, A (sq ft)	Area difference
1	1	4	—	1	—
2	2	8	4	4	3
3	3	12	4	9	5
4	4	16	4	16	7
5	5	20	4	25	9
6	6	24	4	36	11
7	7	28	4	49	13
8	8	32	4	64	15
9	9	36	4	81	17
10	10	40	4	100	19

P- increases by 4 each time
A- increases by 2 more each time

Part D

Explore a different pattern in which squares double in both length and width. Fill in the doubling table. Describe the patterns that you see.

Length (ft)	Width (ft)	Perimeter, P (ft)	Area, A (sq ft)
1	1	4	1
2	2	8	4
4	4	16	16
8	8	32	64
16	16	64	256
32	32	128	1024

P- doubles
A- multiplies by 4

Chapter 13 Performance Task Student Model 2

NAME _____ DATE _____ SCORE _____

Performance Task

Square Deal

Perimeter and area are two ways to measure the size of a figure. In this activity, you will explore patterns of perimeter and area.

Write your answers on another piece of paper. Show all your work to receive full credit.

Part A

Explore a pattern of squares. Create models to show how squares increase in perimeter and area as their side length increases. The first three are done for you. Draw the next seven.

Part B

Now show the numerical patterns of the series. Refer to the models you drew in **Part A** to fill in the table. In words, describe the pattern for both perimeter and area. Which pattern increases in value faster, perimeter or area? Explain.

Length (ft)	Width (ft)	Perimeter, P (ft)	Area, A (sq ft)
1	1	4	1
2	2	8	4
3	3	12	9
4	4	16	16
5	5	20	25
6	6	24	36
7	7	28	49
8	8	32	64
9	9	36	81
10	10	40	100

Area increases faster.

Part C

Now, analyze the patterns for perimeter and area. Find the differences in perimeter and area as the squares increase in side length. Fill in the table. Then describe how each pattern changes.

Length (ft)	Width (ft)	Perimeter, P (ft)	Perimeter difference	Area, A (sq ft)	Area difference
1	1	4	—	1	—
2	2	8	4	4	3
3	3	12	4	9	5
4	4	16	4	16	7
5	5	20	4	25	9
6	6	24	4	36	11
7	7	28	4	49	13
8	8	32	4	64	15
9	9	36	4	81	17
10	10	40	4	100	19

always 4 → odd numbers

Part D

Explore a different pattern in which squares double in both length and width. Fill in the doubling table. Describe the patterns that you see.

Length (ft)	Width (ft)	Perimeter, P (ft)	Area, A (sq ft)
1	1	4	1
2	2	8	4
4	4	16	16
8	8	32	64
16	16	64	256
32	32	128	1024

x2 x4

Chapter 13 Performance Task Student Model 3

NAME _____ **DATE** _____ **SCORE** _____

Performance Task

Square Deal

Perimeter and area are two ways to measure the size of a figure. In this activity, you will explore patterns of perimeter and area.

Write your answers on another piece of paper. Show all your work to receive full credit.

Part A

Explore a pattern of squares. Create models to show how squares increase in perimeter and area as their side length increases. The first three are done for you. Draw the next seven.

Part B

Now show the numerical patterns of the series. Refer to the models you drew in **Part A** to fill in the table. In words, describe the pattern for both perimeter and area. Which pattern increases in value faster, perimeter or area? Explain.

Length (ft)	Width (ft)	Perimeter, P (ft)	Area, A (sq ft)
1	1	4	4
2	2	12	9
3	3	16	16
4	4	20	25
5	5	24	36
6	6	28	49
7	7	32	64
8	8	36	81
9	9	40	100
10	10		

Area

Performance Task (continued)

Part C

Now, analyze the patterns for perimeter and area. Find the differences in perimeter and area as the squares increase in side length. Fill in the table. Then describe how each pattern changes.

Length (ft)	Width (ft)	Perimeter, P (ft)	Perimeter difference	Area, A (sq ft)	Area difference
1	1	4	—	4	—
2	2	8	4	9	3
3	3	12	4	16	5
4	4	16	4	25	7
5	5	20	4	36	9
6	6	24	4	49	13
7	7	28	4	64	15
8	8	32	4	81	17
9	9	36	4	100	19
10	10	40	4		

Part D

Explore a different pattern in which squares double in both length and width. Fill in the doubling table. Describe the patterns that you see.

Length (ft)	Width (ft)	Perimeter, P (ft)	Area, A (sq ft)
1	1	4	4
2	2	8	16
4	4	12	64
8	8	24	256
16	16	40	1024
32	32		

Chapter 13 Performance Task Student Model 4

NAME _____ DATE _____ SCORE _____

Performance Task

Square Deal

Perimeter and area are two ways to measure the size of a figure. In this activity, you will explore patterns of perimeter and area.

Write your answers on another piece of paper. Show all your work to receive full credit.

Part A

Explore a pattern of squares. Create models to show how squares increase in perimeter and area as their side length increases. The first three are done for you. Draw the next seven.

4×4 5×5 6×6 7×7 8×8 9×9 10×10

Part B

Now show the numerical patterns of the series. Refer to the models you drew in **Part A** to fill in the table. In words, describe the pattern for both perimeter and area. Which pattern increases in value faster, perimeter or area? Explain.

Area

Length (ft)	Width (ft)	Perimeter, P (ft)	Area, A (sq ft)
1	1	4	1
2	2		4
3	3		9
4	4		16
5	5		25
6	6		36
7	7		49
8	8		64
9	9		81
10	10		100

Part C

Now, analyze the patterns for perimeter and area. Find the differences in perimeter and area as the squares increase in side length. Fill in the table. Then describe how each pattern changes.

Length (ft)	Width (ft)	Perimeter, P (ft)	Perimeter difference	Area, A (sq ft)	Area difference
1	1	4	—	1	—
2	2		4	4	3
3	3			9	5
4	4			16	7
5	5			25	9
6	6			36	11
7	7			49	13
8	8			64	15
9	9			81	17
10	10			100	19

Part D

Explore a different pattern in which squares double in both length and width. Fill in the doubling table. Describe the patterns that you see.

Length (ft)	Width (ft)	Perimeter, P (ft)	Area, A (sq ft)
1	1	4	1
2	2		4
4	4		16
8	8		64
16	16		256
32	32		1,024

Chapter 14 Performance Task Rubric

Page 163 • Triangle Truth

Task Scenario Students draw a series of precisely defined triangles and use a protractor to measure the angles in those triangles. Students then analyze their data and use it to draw a conclusion—that the sum of the angle measures of any triangle is 180°.		
Depth of Knowledge	\multicolumn{2}{l|}{DOK2, DOK3}	
Part	**Maximum Points**	**Scoring Rubric**
A	2	Full Credit: Students draw all six of the triangles accurately. The equilateral and isosceles triangles should have sides that are equal in length. Right angles, acute angles, and obtuse angles should be drawn accurately. Partial Credit (1 point) will be given for drawing five of the six triangles accurately. No credit will be given for two or more or incorrect triangles.
B	3	Full Credit: Students make accurate angle measurements of all six triangles. Note that four of the six triangles will vary in angle measure. The only triangles whose angle measures can be predicted ahead of time are the equilateral triangle and isosceles right triangle that have measurements of 60°-60°-60° and 90°-45°-45° respectively. Partial Credit (2 points) will be given for making accurate measurements for five of the six triangles. Partial Credit (1 point) will be given for making accurate measurements for three or four of the six triangles. No credit will be given for making inaccurate measurements for three or more triangles.

Grade 4 • **Chapter 14** Performance Task Rubric

Chapter 14 Performance Task Rubric, continued

Part	Maximum Points	Scoring Rubric
C	3	Full Credit: Students correctly determine that the angle measure sum for all six triangles is 180°. (Expect variations of up to 5° in sum totals due to minor inaccuracies in measurement.) Students also correctly draw the conclusion that all triangles have an angle measure sum of 180°. Partial Credit (2 points) will be given if students' data diverges from 180° by more than 5° but is still approximately 180°. Students also correctly draw the conclusion that all triangles have an angle measure sum of 180°. Partial Credit (1 point) will be given if students' data diverges from 180° by more than 5°, or students correctly draw the conclusion that all triangles have an angle measure sum of 180° even if their data is somewhat inaccurate. No credit will be given for failing to obtain an angle measure sum of 180° for two or more of the six triangles, or for failing to make the generalization that all triangles have an angle measure sum of 180°.
D	2	Full Credit: Students formulate a valid hypothesis that states "All triangles have an angle measure sum of 180°." Students also identify valid methods to test their hypothesis, primarily by making measurements for a wide variety of triangles of different sizes and shapes. Partial Credit (1 point) will be given for formulating a valid hypothesis that "All triangles have an angle measure sum of 180°" but failing to identify methods to test their hypothesis. No credit will be given for failing to formulate a valid hypothesis.
TOTAL	10	

Chapter 14 Performance Task Student Model 1

NAME _____ DATE _____

SCORE _____

Performance Task

Triangle Truth

Do all triangles have characteristics in common? In this activity, you will make some discoveries by drawing and measuring triangles and other figures.

Write your answers on another piece of paper. Show all your work to receive full credit.

Part A

Draw triangles 1-6 with the following characteristics:

1. An acute triangle in which all three sides have the same length
2. An acute triangle in which all three sides have different lengths
3. An obtuse triangle in which two sides have the same length
4. An obtuse triangle in which all three sides have different lengths
5. An right triangle in which two sides have the same length
6. An right triangle in which all three sides have different lengths

Part B

Use a protractor to measure the angles of each triangle. Be precise. Record your measurements in the table.

Triangle type	Equal Sides	Angle 1	Angle 2	Angle 3
1	acute	3 equal sides	60	60	60
2	acute	no equal sides	80	48	52
3	obtuse	2 equal sides	140	20	20
4	obtuse	no equal sides	115	37	28
5	right	2 equal sides	90	45	45
6	right	no equal sides	90	30	60

Part C

Find the sum the angle measures for each triangle. Record your calculations. What pattern do you see? Explain.

Triangle type	Equal Sides	Angle 1	Angle 2	Angle 3	Sum
1	acute	3 equal sides	60	60	60	180
2	acute	no equal sides	80	48	52	180
3	obtuse	2 equal sides	140	20	20	180
4	obtuse	no equal sides	115	37	28	180
5	right	2 equal sides	90	45	45	180
6	right	no equal sides	90	30	60	180

All the sums are 180.

Part D

What hypothesis do you make from your data from Part C? How could you find more evidence to support your hypothesis?

The angles of any triangle sum to 180°. I could draw and measure even more triangles.

Chapter 14 Performance Task Student Model 2

NAME _____ DATE _____ SCORE _____

Performance Task

Triangle Truth

Do all triangles have characteristics in common? In this activity, you will make some discoveries by drawing and measuring triangles and other figures.

Write your answers on another piece of paper. Show all your work to receive full credit.

Part A

Draw triangles 1-6 with the following characteristics:

1. An acute triangle in which all three sides have the same length
2. An acute triangle in which all three sides have different lengths
3. An obtuse triangle in which two sides have the same length
4. An obtuse triangle in which all three sides have different lengths
5. A right triangle in which two sides have the same length
6. A right triangle in which all three sides have different lengths

Part B

Use a protractor to measure the angles of each triangle. Be precise. Record your measurements in the table.

Triangle type	Equal Sides	Angle 1	Angle 2	Angle 3
1 acute	3 equal sides	60	60	60
2 acute	no equal sides	60	47	72
3 obtuse	2 equal sides	25	25	130
4 obtuse	no equal sides	127	32	23
5 right	2 equal sides	90	45	45
6 right	no equal sides	90	30	60

Part C

Find the sum the angle measures for each triangle. Record your calculations. What pattern do you see? Explain.

Triangle type	Equal Sides	Angle 1	Angle 2	Angle 3	Sum
1 acute	3 equal sides	60	60	60	180
2 acute	no equal sides	60	47	72	179
3 obtuse	2 equal sides	25	25	130	180
4 obtuse	no equal sides	127	32	23	182
5 right	2 equal sides	90	45	45	180
6 right	no equal sides	90	30	60	180

The sums are all about 180°.

Part D

What hypothesis do you make from your data from Part C? How could you find more evidence to support your hypothesis?

Sum of triangle angle measures is 180°.

Chapter 14 Performance Task Student Model 3

NAME _____ DATE _____

SCORE _____

Performance Task

Triangle Truth

Do all triangles have characteristics in common? In this activity, you will make some discoveries by drawing and measuring triangles and other figures.

Write your answers on another piece of paper. Show all your work to receive full credit.

Part A

Draw triangles 1-6 with the following characteristics:

1. An acute triangle in which all three sides have the same length
2. An acute triangle in which all three sides have different lengths
3. An obtuse triangle in which two sides have the same length
4. An obtuse triangle in which all three sides have different lengths
5. A right triangle in which two sides have the same length
6. A right triangle in which all three sides have different lengths

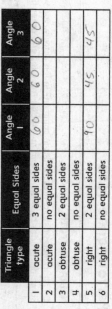

Part B

Use a protractor to measure the angles of each triangle. Be precise. Record your measurements in the table.

Triangle type	Equal Sides	Angle 1	Angle 2	Angle 3
1	acute	3 equal sides	60	60	60
2	acute	no equal sides			
3	obtuse	2 equal sides			
4	obtuse	no equal sides			
5	right	2 equal sides	90	45	45
6	right	no equal sides			

Performance Task (continued)

Part C

Find the sum the angle measures for each triangle. Record your calculations. What pattern do you see? Explain.

Triangle type	Equal Sides	Angle 1	Angle 2	Angle 3	Sum
1	acute	3 equal sides	60	60	60	180
2	acute	no equal sides				180
3	obtuse	2 equal sides				180
4	obtuse	no equal sides				180
5	right	2 equal sides	90	45	45	180
6	right	no equal sides				180

all triangles add to 180

Part D

What hypothesis do you make from your data from Part C? How could you find more evidence to support your hypothesis?

Measure the angles in more triangles.

Chapter 14 Performance Task Student Model 4

NAME _____ DATE _____

SCORE _____

Performance Task

Triangle Truth

Do all triangles have characteristics in common? In this activity, you will make some discoveries by drawing and measuring triangles and other figures.

Write your answers on another piece of paper. Show all your work to receive full credit.

Part A

Draw triangles 1-6 with the following characteristics:

1. An acute triangle in which all three sides have the same length
2. An acute triangle in which all three sides have different lengths
3. An obtuse triangle in which two sides have the same length
4. An obtuse triangle in which all three sides have different lengths
5. A right triangle in which two sides have the same length
6. A right triangle in which all three sides have different lengths

Part B

Use a protractor to measure the angles of each triangle. Be precise. Record your measurements in the table.

Triangle type	Equal Sides	Angle 1	Angle 2	Angle 3
1	acute	3 equal sides	60	60	60
2	acute	no equal sides	50	60	70
3	obtuse	2 equal sides	130	25	25
4	obtuse	no equal sides	130	40	10
5	right	2 equal sides	90	45	45
6	right	no equal sides	90	50	40

Performance Task (continued)

Part C

Find the sum the angle measures for each triangle. Record your calculations. What pattern do you see? Explain.

Triangle type	Equal Sides	Angle 1	Angle 2	Angle 3	Sum
1	acute	3 equal sides	60	60	60	180
2	acute	no equal sides	50	60	70	180
3	obtuse	2 equal sides	130	25	25	180
4	obtuse	no equal sides	130	40	10	180
5	right	2 equal sides	90	45	45	180
6	right	no equal sides	90	50	40	180

I know that the angles always add to 180.

Part D

What hypothesis do you make from your data from Part C? How could you find more evidence to support your hypothesis?

Angles of any triangle total 180.

Benchmark Test 1

NAME _____ DATE _____ SCORE _____

Benchmark Test 1

1. The Garcias are buying a car priced between $10,000 and $20,000. The car's price is a multiple of 100. The hundreds digit is 5 times the thousands digit.

   Part A: Write the price of the car in word form.

   eleven thousand, five hundred

   Part B: Fill in the blanks to complete the expansion of the car's price.

   [1] × 10,000 + [1] × 1,000 + [5] × 100 + [0] × 10 + [0] × 1

2. The table shows the numbers of 4 different types of CDs sold at a chain of music stores. Select > or < to compare the indicated numbers.

Type of CD	Number Sold
Pop	20,756
Country	19,984
Rock	21,001
Hip Hop	20,689

   > < the number of country CDs _____ the number of hip hop CDs

   ☐ ☐ the number of pop CDs _____ 20,000

   ☐ ☐ the number of rock CDs _____ the number of hip hop CDs

   ☐ ☐ the number of country CDs _____ 20,000

   ☐ ☐ the number of hip hop CDs _____ the number of pop CDs

   ☐ ☐ the number of rock CDs _____ the number of pop CDs

3. Rashan has $23,456 in his college savings account. Joannie has $14,365 in her college savings account. Sort the digits below according to their value in each student's account. Some digits may not be used.

   | 1 | 2 | 3 | 4 | 5 | 6 |

Value of Digits	
Worth 10 times more in Rashan's account than in Joannie's	Worth 10 times more in Joannie's account than in Rashan's
3, 5	4, 6

---

4. The digit in the hundred thousands place is 4 times the digit in the ones place. The digit in the tens place is 3 times the digit in the ones place. The digit in the thousands place has 10 times the value of the digit in the hundreds place. The digit in the ten thousands place has 10 times the value of the digit in the thousands place. Fill in the missing digits to make the inequality statement true.

   [8] [3] [3,] [3] [6] [ ] [2] > 500,000

5. Andrea agreed to sell 100 raffle tickets for a band fundraiser. The table shows the number of tickets she sold during each of the first three weeks of the month. She estimates that she has about 30 more to sell. Use the numbers to write an equation that shows how Andrea could have come up with her estimate. Some numbers may not be used.

Week	Number of Tickets Sold
1	21
2	8
3	37

   0   10   20   30   40   50   60   70   80   100

   100 − (20 + 10 + 40) = 30

6. Tariq is saving money for a trip to an amusement park. He will need $50 for the ticket and $20 for food. Tariq saves the $42 he earned helping at his uncle's business. His sister gives him a coupon for $10 off admission to the park. Select all expressions that are equal to the remaining amount, in dollars, that Tariq must save.

   ■ 50 + 20 − (42 + 10)   ☐ (50 + 20) − (42 + 10)

   ☐ (50 + 20) − 42 + 10   ☐ (50 + 20) − 42 − 10

   ☐ 50 + 20 − 42 − 10   ■ (50 + 20) − (42 − 10)

7. Complete the multiplication problem by writing in the missing digits.

Benchmark Test 1

11. A gift shop owner has 21 candles. She puts the same number in each of 7 gift baskets.

    Part A: Write a division statement whose answer is the number of candles in each basket. Include the answer in your statement.

    $21 \div 7 = 3$

    Part B: Represent the division statement on a number line.

12. A farmer arranged some pumpkins as shown. Show 2 more arrangements that use the same total number of pumpkins. Within an arrangement, each row should have the same number of pumpkins.

---

8. Martina and Barry rented a boat. They filled it with 18 gallons of gas. They used 6 gallons of gas to get to an island and 3 gallons waterskiing and sightseeing.

    Part A: Write an equation that can be used to find g, the number of gallons of gas that are left. Use the numbers and variable shown to fill in the boxes.

    | 3 | 6 | 18 | g |

    $18 - (\boxed{6} + \boxed{3}) = \boxed{g}$

    Part B: Solve the equation.

    $18 - (6 + 3) = g$
    $18 - 9 = g$
    $9 = g$; 9 gallons

9. The school dance took in $1,100. Expenses totaled $304. The dance advisor wrote a subtraction problem to figure out the amount of profit from the dance, after expenses. Select whether each statement describes a step that can be used to find the answer to the advisor's subtraction problem.

Yes	No	
☐	☐	Subtract 0 tens from 0 tens, which leaves 0 tens.
☐	☐	Subtract 0 thousands from 1 thousand, which leaves 1 thousand.
☐	☐	Use one hundred as 10 tens.
☐	☐	Use one ten as 10 ones, which leaves 0 tens and 10 ones.
☐	☐	Subtract 4 ones from 10 tens.
☐	☐	Use one ten as 10 ones, which leaves 9 tens and 10 ones.

10. Ms. Albright has 42 packages of paper. Each package has 50 sheets of paper. Mrs. Albright wants to find the total number of sheets of paper she has. Which is the best number sentence for her to use as a first step?

    A. $42 \times 5 = 210$ C. $42 + 50 = 92$

    B. $42 + 5 = 47$ D. $42 \times 15 = 630$

Benchmark Test 1

13. Alice has 8 dimes. Noah has 24 dimes. The equation 24 = 8 × 3 can be used to describe the relationship between the number of dimes that Alice and Noah have. Which statement below also represents this relationship?

   A. Noah has 3 more dimes than Alice.
   B. Noah has 3 times as many dimes as Alice.
   C. Alice has 3 more dimes than Noah.
   D. Alice has 3 times as many dimes as Noah.

14. Hitomi, Julius, and five of their friends each have the same number of crayons. The total number of crayons they have is shown. Circle groups of crayons to model the number the each student has.

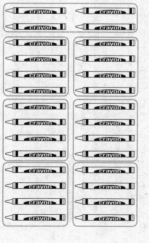

15. Look at the area model. Complete the number sentence so that it shows the partial products and final product shown by the model. Fill in the missing numbers.

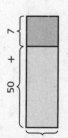

(50 × 4) + (7 × [4]) = [57] × 4 = [228]

16. A catering company will earn the same amount on each of the 3 days of an event. To figure out about how much it would earn, the company estimated the product ___ 82 × 3 as 2,400 by first rounding to the greatest place value possible.

   Part A: What is the value of the missing digit?

   [7]

   Part B: Is the estimate greater or less than the actual number? Explain.

   Sample answer: The estimate is greater. Because of the 8 in the tens place, the number with the missing digit would be rounded up, which makes the estimate greater than the actual number.

17. Select **all** of the problems that would require regrouping to find the product.

   ☐ 322 × 5   ☐ 241 × 4   ☐ 3,523 × 2   ☐ 31,312 × 3   ☐ 4,523 × 6

18. Desiree multiplied 3,492 by 5 and got 174,600. Use place value to explain the mistake that she made.

   Sample answer: The answer has an extra zero at the end. When she multiplied 2 ones by 5 ones, she got 10 and then wrote down 2 zeros instead of one. She might have thought she was multiplying by tens, not ones.

Benchmark Test 1

19. A building has *f* floors. Each floor has 34 offices. The number *f* is an odd 2-digit number. Indicate whether each statement is true or false.

True	False	
☐	☐	The total number of offices in the building could be a multiple of 10.
☐	☐	The number of floors must be a multiple of 5.
☐	☐	The total number of offices in the building is greater than 350.
☐	☐	The total number of offices in the building is divisible by *f*.
☐	☐	The number equal to 34 × *f* could be greater than 3,400.

20. A company ordered 23 sandwich platters and 32 salads for a meeting. Sandwich platters cost $13 each, and salads cost $7 each. The company paid a $6 delivery fee.

Part A: Use some of the numbers and symbols shown to complete the equation so that it shows the total amount, *n*, that the company paid for the order.

| 6 | 7 | 13 | 23 | 32 | + | × | ÷ | − |

23 × 13 + 32 × 7 + 6 = *n*

Part B: Solve the equation. Show your work and state the answer.

23 × 13 + 32 × 7 + 6 = *n*
299 + 224 + 6 = *n*
299 + 230 = *n*
529 = *n*
The company paid $529.

Benchmark 1 Performance Task Rubric

Page 172 • Giant Trees

Task Scenario
Students will find data on giant sequoia trees and use it to compare multi-digit numbers using inequality symbols and by dividing to find how many times greater one number is than another.

Depth of Knowledge	DOK2, DOK3

Part	Maximum Points	Scoring Rubric
A	4	Full Credit:

Tree Name	Height (feet)	Trunk Volume (cubic feet)
Boole	269	42,472
General Grant	267	46,608
General Sherman	248	52,500
Grizzly Giant	209	34,005
Hart	278	34,407

Volume (in Cubic Feet) of the World's Largest Sequoias			
< 40,000	> 40,000 < 45,000	> 45,000 < 50,000	> 50,000
Grizzly Giant, Hart	Boole	General Grant	General Sherman

Partial Credit (3 points) will be given if the first table is incorrect or if there is one mistake in the second table.

Partial Credit (2 points) will be given if there are two mistakes in the second table or if there is one mistake in the second table and the first table is incorrect.

Partial Credit (1 point) will be given if there are three mistakes in the second table or if there are two mistakes in the second table and the first table is incorrect.

No credit will be given if the second table is not filled out/completely incorrect AND the first table is not filled out.

Benchmark 1 Performance Task Rubric, continued

Part	Maximum Points	Scoring Rubric
B	2	Full Credit: $y = 52{,}500 \div 5$ (or equivalent) $y = 10{,}500$ cubic feet Partial Credit (1 point) will be given if the quotient is incorrect but the equation is correct. No credit will be given for a correct quotient without an equation or work shown.
TOTAL	6	

Benchmark Test 2

NAME _____ DATE _____ SCORE _____

1. A bookshop owner must place 72 books equally on 6 different shelves. The model below shows how he did this.

   Part A: Complete the division sentence shown by the model.

   $72 \div 6 = 12$

   Part B: What does the answer to the division sentence mean in terms of the problem situation?

   The owner places 12 books on each shelf.

2. Look at the number list.

   Part A: Extend the pattern.

   27, 30, 21, 24, 15, 18, __9__, __12__

   Part B: Complete the statement.

   The pattern is _add 3, subtract 9_.

3. Jocelyn has $\frac{4}{6}$ cup of lemon juice. She adds an additional $\frac{3}{9}$ cup to it.

   Part A: Write each fraction in lowest terms.

   $\frac{2}{3}, \frac{1}{3}$

   Part B: How much lemon juice does she have in all?

   1 cup

4. An electrician has 5 lengths of wire. The table shows the lengths in yards.

Wire	A	B	C	D	E
Length (yd)	$\frac{1}{4}$	$\frac{1}{16}$	$\frac{5}{16}$	$\frac{1}{8}$	$\frac{3}{8}$
Rank	3	1	4	2	5

   Part A: Rank the wires from least (1) to greatest (5) on the table.

   Part B: Where would a wire that is $\frac{5}{8}$ yard long rank?

   All the way at the end, at rank 6.

5. Mrs. Johansen has two gallon-sized containers of milk. The first one has $\frac{5}{6}$ gallon in it. When she pours the second container of milk into the first, she gets exactly 1 gallon. How many gallons were in the second container?

   $\frac{1}{6}$ gallon

6. Mark "Yes" for each division statement that has a remainder and "No" for each that does not.

Yes	No	
☐	☐	120 ÷ 10
☐	☐	66 ÷ 5
☐	☐	131 ÷ 4
☐	☐	63 ÷ 7
☐	☐	132 ÷ 11

Benchmark Test 2

7. Giovanni conducted a survey of how far people drive to work. The results are shown in the table.

Town	Drive to Work More than 10 miles	Drive to Work Less than 10 miles
Springfield	15	30
Jacksontown	22	18
Blakesville	10	40

Part A: The ratio of drivers who drive more than 10 miles to those who drive less than 10 miles for the town of Clarksville is the same as for the town of Blakesville. If 20 people surveyed in Clarksville drive more than 10 miles to work, how many people in the Clarksville survey drive less than 10 miles to work?

80

Part B: Which town had more than half of their drivers driving more than 10 miles to work?

Jacksontown

8. Janice listed the first 6 multiples of 5 on the board. Select all correct statements.

☐ The ones digit in all of the numbers is 5.
☐ After 5, the tens digit follows the pattern 1, 1, 2, 2, 3, 3, ...
☐ The sum of the digits in each number in the list is divisible by 5.
☐ Every other number ends in 0.
☐ Each number in the list is divisible by 5.

9. Franco bought a notepad for $0.85. Circle the smallest number of coins that Franco can use to pay for the notepad.

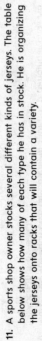

10. A scientist weighed a lizard specimen and found it to weigh 0.5 ounces. Circle all the numbers that have the same value as 0.5.

0.50 $\frac{5}{10}$ $\frac{5}{100}$ $\frac{50}{100}$ $\frac{1}{2}$

11. A sports shop owner stocks several different kinds of jerseys. The table below shows how many of each type he has in stock. He is organizing the jerseys onto racks that will contain a variety.

Type of Jersey	Number of Jerseys
Football	47
Soccer	38
Baseball	41
Volleyball	32

Part A: On one set of racks he wants to hang the football and soccer jerseys. Each rack can hold 9 shirts. How many racks does the owner need? What is the significance of the remainder?

10 racks are needed. 85 ÷ 9 = 9 R4. 9 racks will be full, and the tenth rack will have the remainder (4 jerseys).

Part B: The owner has eight racks to dedicate to the baseball and volleyball jerseys. Can he fit them all? If not, how many will be left over?

No; 73 ÷ 9 = 8 R1, so 1 shirt will be left without a rack.

Benchmark Test 2

12. A stamp collector buys 7 stamps during his first year of collecting stamps. After the first year, he buys 6 more stamps than he did the previous year.

    Part A: How many stamps did he purchase in the 5th year?

    31 stamps

    Part B: Find the total number of stamps that he purchases in the first 5 years. Show your work.

    95 stamps; 7 + 13 + 19 + 25 + 31 = 95

13. Write an addition sentence for each model. Then find the sum.

    Part A:

    $\frac{3}{8} + \frac{5}{8} = 1$

    Part B:

    $\frac{5}{12} + \frac{9}{12} = 1\frac{2}{12} = 1\frac{1}{6}$

    Part C:

    $1\frac{1}{4} + 1\frac{3}{4} = 3\frac{1}{2}$

14. A painter has two containers of paint. The first container has 0.67 gallons in it. The second has 0.3 gallons in it. Shade the models below and find the total amount of paint the painter has.

    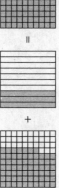

    0.97 gallons

15. A florist is trying to split 36 flowers in vases with the following criteria.
    1. Each vase must have more than 2 flowers.
    2. Each vase must have the same number of flowers.

    How many different arrangements of vases can the florist form? List them all in the box below.

    2 vases of 18 flowers
    3 vases of 12 flowers
    4 vases of 9 flowers
    6 vases of 6 flowers
    9 vases of 4 flowers
    12 vases of 3 flowers
    There are 6 ways.

16. Theo has raised $819 for charity. He wants to donate the same amount of money to each of nine charities. He uses the division statement $810 ÷ 9 to estimate that he will give about $90 to each charity. Is this a reasonable estimate? Explain.

    Sample answer: Yes, this is a reasonable estimate. $819 is close to $810, which is a convenient number for division by 9.

Benchmark Test 2

17. Look at the table.

Input (r)	10	12	15	17	20	25
Output (v)	15	17	20			

Which equation describes the pattern?

A. $r \times 5 = v$
B. $v \times 5 = r$
C. $r + 5 = v$ ⬅
D. $v + 5 = r$

18. Determine whether each statement is true or false.

	True	False
$\frac{3}{9}$ is equivalent to $\frac{1}{3}$	■	☐
$\frac{3}{6}$ is the simplest form of $\frac{12}{24}$	☐	☐
$1\frac{8}{12}$ is equivalent to $\frac{5}{3}$	■	☐
$\frac{7}{16}$ is the simplest form of $\frac{14}{32}$	☐	☐

19. Enrique rides his bike $\frac{3}{4}$ mile to school every day and the same $\frac{3}{4}$ mile home. If he has school five days each week, circle all of the expressions that will find the total number of miles Enrique rides every week.

$5 \times \frac{3}{4}$ $5 \times \frac{6}{4}$ $5 \times \frac{3}{2}$

$10 \times \frac{3}{4}$ $10 \times \frac{3}{2}$ $\frac{6}{4} + \frac{6}{4} + \frac{6}{4} + \frac{6}{4} + \frac{6}{4}$

20. A grocer unpacks boxes of bananas that weigh 87.24 pounds altogether. He started working on packaging the bananas at 9:30 A.M. When he finished filling the bin with all the bananas that would fit, there were 6.21 pounds of bananas left over. With this information, which of the following can you solve? Select all that apply.

 How many pounds of bananas fit in the bin?
☐ When did the grocer finish his work?
☐ How long did it take the grocer to put out the bananas?
☐ How many boxes of bananas did he unpack?

Benchmark 2 Performance Task Rubric

Page 180 • Selling Tomatoes

Task Scenario
Students will use addition of fractions and multiplication of a fraction by a whole number to figure out how many bushels and pecks of tomatoes a farmer sells. Students will need to research how many pecks are in a bushel.

Depth of Knowledge	DOK2, DOK3, DOK4

Part	Maximum Points	Scoring Rubric
A	2	Full Credit: Student correctly identifies that there are 4 pecks in a bushel AND that a peck is one fourth of a bushel. Partial Credit (1 point) will be given for identifying 4 pecks in a bushel **OR** that a peck is one fourth of a bushel. No credit will be given for two incorrect answers.
B	3	Full Credit: $33 \times \frac{1}{4} = 8\frac{1}{4}$ Partial Credit (1 point) will be given for each entry in the number sentence. No credit will be given for three incorrect entries.
C	2	Full Credit: $6; Student correctly shows work of $24 \times \frac{1}{4} = 6$. Partial Credit (1 point) will be given for a correct answer of $6 without proper work. No credit will be given for an incorrect answer.
D	2	Full Credit: Student correctly identifies the morning as the higher selling period because $5\frac{1}{4} > 4\frac{3}{4}$. Student correctly compares both periods using the same unit. Partial Credit (1 point) will be given for a correct answer (morning) without a proper explanation. No credit will be given for an incorrect answer.
TOTAL	9	

NAME _____ DATE _____

SCORE _____

Benchmark Test 3

1. Order the numbers from *greatest to least*.

   654,879 _____

   654,978 _____

   645,879 _____

   645,789 _____

2. Circle the statement that shows *four hundred sixty-four thousand, five hundred four* in expanded form.

   A. 400,000 + 40,000 + 6,000 + 500 + 40

   B. 464,504

   C. 446,450

   D. (400,000 + 60,000 + 4,000 + 500 + 4)

3. List all of the factors of 24.

   1, 2, 3, 4, 6, 8, 12, 24

4. *Part A:* Find the unknown.

   350 + (25 + __19__) = (350 + 25) + 19

   Part B: Circle the addition property that is modeled.

   A. (Associative Property of Addition)

   B. Commutative Property of Addition

   C. Identity Property of Addition

   D. Parentheses Property of Addition

5. Subtract. Check using estimation.

   ```
     608,256          608,000
   − 165,704        − 166,000
   ─────────        ─────────
     442,552   Sample estimate:  442,000
   ```

6. Use multiplication to complete the number sentence. 3 times as many

   3 × 4 = 12

7. List the first five multiples of 11. Explain how you know what the first five multiples are?

   11, 22, 33, 44, 55

   Sample answer: You find the first five multiples of a number by multiplying the number by 1, 2, 3, 4, and 5.

Benchmark Test 3

8. Use the clues to complete the place-value chart.

Thousands			Ones		
hundreds	tens	ones	hundreds	tens	ones
7	4	1	2	9	6

The 1 has a value of 1 × 1,000.
The 2 has a value of 2 × 100.
The 4 is in the ten thousands place.
The 6 is in the ones place.
The 7 has a value of 7 × 100,000.
The 9 is in the tens place.

9. Vincenzo joined a basketball league. It costs $23 a week. If the league lasts eight weeks, how much does Vincenzo pay? Show your work.

$23 × 8 = $184

Sample work:
```
    2
  $23
  ×  8
 $184
```

10. Circle the correct answer.
2,272 × 4 =
A. 1,568
B. **9,088**
C. 8,888
D. 8,088

11. Part A: Estimate.

Sample estimate shown.

```
  $71  →   $70
  × 52 →   × 50
          3,500
```

Part B: Solve the problem. Is the actual answer *greater than* or *less than* the estimate? Explain.

$3,692. The answer is greater than the estimate. Sample answer: Since both numbers were rounded down, the product is less than the actual answer.

12. There are 126 pencils in a box. Each classroom in the school gets a box. There are 40 classrooms. How many pencils are there?

Part A: Circle all the correct choices that show a correct way to solve the problem.

A. 126 × 40 =
B. 126
    × 40
C. 40
    × 126
D. **40 × 126 =**

Part B: Solve the problem.

5,040 pencils

Benchmark Test 3

13. Divide. Use multiplication and addition to check.

```
   730 R1
6)4,381
  -42
   18
  -18
    01
```

$$\begin{array}{r}\$730\\\times 6\\\hline 4,380\end{array}\qquad\begin{array}{r}4,380\\+1\\\hline 4,381\end{array}$$

14. Divide. Use the Distributive Property and the area model to show your work.

425 ÷ 5 =

	80	4	1
5	400	20	5

80 ÷ + 4 ÷ + 1 ÷ = 85

425 ÷ 5 = 85

15. Extend the pattern.

16. Look at the table below.

    Part A: Complete the table.

Input (a)	2	5	8	11	14
Output (z)	14	17	20	23	26

    Part B: Circle the equation that describes the pattern.

    A. $a + 12 = z$

    B. $z + 12 = a$

    C. $a \times 7 = z$

    D. $a + 3 = z$

17. Circle all of the fractions that make the equation true when substituted for x.

    $$\frac{5}{10} = x$$

    A. $\frac{1}{2}$

    B. $\frac{3}{6}$

    C. $\frac{4}{10}$

    D. $\frac{5}{8}$

Benchmark Test 3

18. Use the rectangle below to answer the following questions.

    Part A: Shade $\frac{3}{5}$ of the rectangle below.

    Part B: Write an equivalent fraction.

    Sample answer: $\frac{6}{10}$

19. Tai and her four friends cut some honeydew melons into thirds.

    Part A: If each girl takes $\frac{1}{3}$ of a melon, how much melon would they have taken together? Explain by writing an equation.

    $\frac{5}{3}$ or $1\frac{2}{3}$ melons;
    Sample equation: $\frac{1}{3}+\frac{1}{3}+\frac{1}{3}+\frac{1}{3}+\frac{1}{3}=\frac{5}{3}$ or $1\frac{2}{3}$

    Part B: If the girls cut 3 melons, how many thirds were left over after they each took their share? Explain by writing an equation.

    $\frac{4}{3}$ or $1\frac{1}{3}$ melons; $3\times\frac{3}{3}=\frac{9}{3}-\frac{5}{3}=\frac{4}{3}$ or $1\frac{1}{3}$ melons left

    Part C: Place a point on the number line for the fraction that represents the total amount of melon taken.

    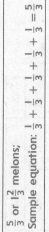

20. Circle all of the fractions that are correct.

    $\frac{5}{8} + \frac{7}{8} =$

    A. $1\frac{1}{4}$
    B. $\frac{3}{2}$
    C. $1\frac{4}{8}$
    D. $\frac{6}{4}$

21. Emerson has $1\frac{5}{6}$ oranges. She eats $\frac{7}{6}$ oranges. Does she have $\frac{2}{3}$ orange left? Circle yes or no.

    Yes   No

22. Shade the grid model to show forty-eight out of a hundred.

    Part A: Shade the model.

    Part B: Write forty-eight out of a hundred as a decimal.

    0.48

    Part C: Write forty-eight out of a hundred as a fraction in simplest form.

    $\frac{12}{25}$

23. Landon rode 0.87 miles on his scooter. Piper rode 0.78 miles on her bike. Who traveled farther? Explain.

    Landon; 0.87 > 0.78

24. Jaylen spends four hours hiking. How many minutes does he hike? Write an equation.

    240 minutes; 60 × 4 = 240

Benchmark Test 3

25. Complete the conversion table.

Pints (pt)	Gallons (gal)	(pt, gal)
16	2	(16, 2)
32	4	(32, 4)
48	6	(48, 6)
64	8	(64, 8)

26. Complete the conversion table.

Centimeters (cm)	Meters (m)	(cm, m)
100	1	(100, 1)
400	4	(400, 4)
600	6	(600, 6)
800	8	(800, 8)

27. Choose the best estimate for the length of a surfboard.

A. 3 meters ⬚
B. 6 meters
C. 3 kilometers
D. 6 centimeters

28. Angel is building a sandbox. It will be 3 feet wide and 4 feet long.

Part A: What is the perimeter of the sandbox? Explain.

3 ft + 3 ft + 4 ft + 4 ft = 14 ft.

Part B: What is the area of the sandbox? Explain.

3 ft × 4 ft = 12 sq ft.

29. Rian draws an angle with a measure greater than 90° and less than 180°. What type of angle did she draw?

A. acute
B. obtuse ⬚
C. perpendicular
D. right

30. Draw the lines of symmetry in the figure.

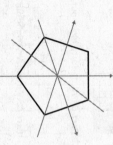

31. Cydney's school is on Main Street, which runs between 4th Avenue and 5th Avenue. Both 4th Avenue and 5th Avenue intersect Main Street at right angles. How would you describe the lines represented by 4th Avenue and 5th Avenue?

A. Intersecting lines
B. Perpendicular lines
C. Obtuse lines
D. Parallel lines ⬚

Benchmark 3 Performance Task Rubric

Page 191 • Organizing a Collection

Task Scenario
Students will use a data set to create a line plot, analyze data from the line plot, subtract fractions, convert decimals to fractions, and compare decimals.

Depth of Knowledge	DOK1, DOK2, DOK3	
Part	**Maximum Points**	**Scoring Rubric**
A	4	**Full Credit:** The student correctly creates a line plot from the data given, including all labels for the number line and a title. **Rock and Mineral Length** [Line plot showing X marks above points on number line from 0 to 3 in quarter-inch increments, labeled "Inches"] Partial Credit (3 points) will be given for creating a line plot from the data given but omitting the labels for the number line and title **OR** correctly creating the line plot including labels for the number line and title, but incorrectly plotting one to five data points. Partial Credit (2 points) will be given for creating a line plot with 10–14 of the data points plotted correctly but omitting the labels for the number line and title **OR** correctly creating the line plot including labels for the number line and title, but incorrectly plotting 6–10 data points. Partial Credit (1 point) will be given for creating a line plot with 6–10 of the data points plotted correctly but omitting the labels and for the number line title **OR** correctly creating the line plot including labels for the number line and title, but incorrectly plotting 0–5 data points. No credit will be given for an incorrect answer.

Benchmark 3 Performance Task Rubric, continued

Part	Maximum Points	Scoring Rubric
B	3	**Full credit:** The student correctly identifies the length of the largest rock ($2\frac{1}{2}$ inches), the length of the smallest rock ($\frac{1}{2}$ inch), and the highest frequency measurement (2 inches). **Partial Credit (2 points)** will be given if the student correctly identifies two of the following: the length of the largest rock ($2\frac{1}{2}$ inches), the length of the smallest rock ($\frac{1}{2}$ inch), and the highest frequency measurement (2 inches). **Partial Credit (1 point)** will be given if the student correctly identifies one of the following: the length of the largest rock ($2\frac{1}{2}$ inches), the length of the smallest rock ($\frac{1}{2}$ inch), and the highest frequency measurement (2 inches). No credit will be given for an incorrect answer.

Part	Maximum Points	Scoring Rubric
C	2	**Full Credit:** The student correctly identifies the difference between the length of the longest and shortest rocks (2 inches) and explains how the answer was obtained. **Partial credit (1 point)** will be given for correctly identifying the difference between the length of the longest and shortest rocks. No credit will be given for an incorrect answer.
D	2	**Full Credit:** The student correctly converts 0.42 to $\frac{42}{100}$ and correctly locates it on a number line. **Partial Credit (1 point)** The student correctly converts 0.42 to $\frac{42}{100}$ **OR** correctly locates it on a number line. No credit will be given for an incorrect answer.
E	3	**Full Credit:** The student correctly locates 0.42 and 0.29 on the number line and correctly writes the comparison 0.42 > 0.29. **Partial credit (2 points)** will be given for correctly locating one number on the number line and correctly writing the comparison **OR** correctly locating both numbers on the number line but omitting or incorrectly writing the comparison. **Partial credit (1 point)** will be given for correctly locating one number on the number line but omitting or incorrectly writing the comparison **OR** for omitting or incorrectly placing the numbers on the number line but correctly writing the comparison. No credit will be given for an incorrect answer.
TOTAL	14	

Benchmark Test 4

NAME _____ DATE _____ SCORE _____

Benchmark Test 4

1. Use the clues to complete the place-value chart.

Thousands			Ones		
hundreds	tens	ones	hundreds	tens	ones
8	9	5	1	7	3

The 1 has a value of 1 × 100.

The 3 has a value of 3 × 1.

The 5 is in the thousands place.

The 7 is in the tens place.

The 8 has a value of 8 × 100,000.

The 9 is in the ten thousands place.

2. Angelica joined a golf league. It costs $37 per week. If the league lasts six weeks, how much does Angelica pay? Show your work.

$37 × 6 = $222

3. Circle the correct answer.

1,598 × 5 =

A. 7,990
B. 3,950
C. 6,990
D. 7,450

---

4. 572 + (84 + _____) = (572 + 84) + 65

Part A: Find the unknown.

65

Part B: Circle the addition property that is modeled.

A. Associative Property of Addition
B. Commutative Property of Addition
C. Identity Property of Addition
D. Parentheses Property of Addition

5. Subtract. Check using estimation.

```
  984,065
- 654,876
  -------
  329,189
```

```
  1,000,000
-   700,000
  ---------
    300,000
```

6. Use multiplication to complete the number sentence.

3 × 5 = 15

7. List the first six multiples of 12. Explain how you know what the first six multiples are.

12, 24, 36, 48, 60, 72;
Sample answer: You find the first six multiples of a number by multiplying the number by 1, 2, 3, 4, 5, and 6.

Benchmark Test 4

8. List all the factors of 48.

   1, 2, 3, 4, 6, 8, 12, 16, 24, 48

9. Order the numbers from *greatest to least*.

   835,468   853,864

   853,846   853,846

   835,846   835,846

   853,864   835,468

10. Circle the statement that shows *six hundred five thousand, two hundred eleven* in expanded form.

    A. 650,211

    B. 605,211

    C. 600,000 + 50,000 + 200 + 10 + 1

    D. 600,000 + 5,000 + 200 + 10 + 1

---

11. Jaden draws an angle with a measure greater than 0° and less than 90°. What type of angle did he draw?

    A. acute

    B. obtuse

    C. perpendicular

    D. right

12. Draw the lines of symmetry in the figure.

    ONLINE TESTING
    On the test, you may be asked to draw lines of symmetry on a figure in a grid using a computer. In this book, you will draw the lines by hand.

13. Kai's favorite library is on State Street, which runs between Ohio Avenue and Michigan Road. If State Street intersects Ohio Avenue at an 80° angle and Michigan Road at a right angle, how would you describe the lines that State Street and Michigan Road make?

    A. acute lines

    B. perpendicular lines

    C. obtuse lines

    D. parallel lines

Benchmark Test 4

14. Look at the table below.

    Part A: Complete the table.

Input (b)	Output (g)
3	6
4	8
7	14
17	34
41	82

    Part B: Circle the equation that describes the pattern.

    A. $b + 5 = g$
    B. $g + 5 = b$
    C. $b + 7 = g$
    D. $b \times 2 = g$ ⟵ (circled)

15. Circle all of the fractions that make the equation true when substituted for x.

    $$\frac{3}{9} = x$$

    A. $\frac{2}{3}$
    B. $\frac{1}{3}$ (circled)
    C. $\frac{4}{12}$ (circled)
    D. $\frac{1}{6}$

16. Kenzie walks 0.65 mile to the soccer fields. Armani walks 0.58 miles to the soccer fields. Who walks farther? Explain.

    Kenzie; 0.65 > 0.58

17. **Part A:** Estimate.

    ```
       79  →   80
     × 37  → × 40
            ─────
             3,200
    ```

    Part B: Solve the problem. Is the actual answer *greater than* or *less than* the estimate? Explain.

    2,923; The estimate is greater than the answer. Sample answer: The estimate is greater because both numbers were rounded up.

18. There are 150 newspapers in a bundle. Each neighborhood or apartment building gets a bundle. There are 22 neighborhoods and apartment buildings. How many newspapers are there?

    Part A: Circle all the choices that show a correct way to solve the problem.

    A. $150 \times 22 =$ (circled)
    B. $150 \times 20 + 150 \times 2 =$ (circled)
    C. $150 + 20 \times 150 + 2 =$
    D. $22 \times 150 =$ (circled)

    Part B: Solve the problem.

    3,300

    ONLINE TESTING
    On the actual test, you may have to click on multiple bubbles. In this book, you will circle the answers.

Benchmark Test 4

19. Divide. Show your work. Use multiplication and addition to check.

```
    725
8)5,802
   -56
    20
   -16
    42
   -40
     2
```

```
   725
  ×  8
 5,800
```

```
 5,800
 +   2
 5,802
```

20. Divide. Use the Distributive Property and the area model to show your work.

$372 \div 6 =$

```
        50      10   2
      ┌────────┬────┬────┐
   6  │  300   │ 60 │ 12 │
      └────────┴────┴────┘
```

$50 + 10 + 2 = 62$

$372 \div 6 = 62$

21. Draw the next figure in the pattern.

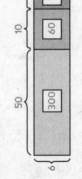

22. Complete the conversion table.

Cups (c.)	Quarts (qt.)	(c., qt.)
4	1	(4, 1)
16	4	(16, 4)
24	6	(24, 6)
28	7	(28, 7)

23. Complete the conversion table.

Centimeters (cm)	Kilometers (km)	(cm, km)
100,000	1	(100,000, 1)
200,000	2	(200,000, 2)
500,000	5	(500,000, 5)
800,000	8	(800,000, 8)

24. Choose the best estimate for the height of a door.

A. 3 centimeters
B. 6 kilometers
C. 3 meters (circled)
D. 6 meters

25. Dante is putting a fence around his garden. The garden will be 2 feet wide and 8 feet long.

Part A: What is the perimeter of the garden? Explain.

2 ft. + 8 ft. + 2 ft. + 8 ft. = 20 ft.

Part B: What is the area of the garden? Explain.

2 ft. × 8 ft. = 16 sq. ft.

Benchmark Test 4

26. Cole and six of his friends cut some grapefruits into halves.

   Part A: If each person takes $\frac{1}{2}$ of a grapefruit, how many grapefruits would they have taken together? Explain by writing an equation.

   $\frac{7}{2}$ or $3\frac{1}{2}$ grapefruits;
   Sample answer: $\frac{1}{2}+\frac{1}{2}+\frac{1}{2}+\frac{1}{2}+\frac{1}{2}+\frac{1}{2}+\frac{1}{2}=\frac{7}{2}$

   Part B: If the friends cut 5 grapefruits, how many halves were left over after they each took their shares? Explain by writing equations.

   $\frac{3}{2}$ or $1\frac{1}{2}$ grapefruits;
   Sample answer: $5\times\frac{2}{2}=\frac{10}{2},\ \frac{10}{2}-\frac{7}{2}=\frac{3}{2}$

   Part C: Place a point on the number line for the fraction that represents the total amount of grapefruit taken.

   ONLINE TESTING
   On the test, you might click on the number line to answer the question. In this book, you will write on the number line.

27. Look at the rectangle below.

   Part A: Shade $\frac{5}{8}$ of the rectangle.

   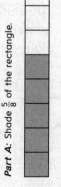

   Part B: Write an equivalent fraction.

   $\frac{10}{16}$

28. Circle all of the fractions that are equal to the sum.

   $\frac{5}{6}+\frac{4}{6}=$

   A. $\frac{3}{2}$
   B. $1\frac{1}{3}$
   C. $2\frac{1}{6}$
   D. $\frac{9}{6}$

29. Muhammad had 2 carrots. He ate $\frac{7}{4}$ carrots. Does he have $\frac{1}{2}$ carrot left? Circle yes or no.

   Yes   No

30. Shade the grid model to show seventy-two out of a hundred.

   Part A: Shade the model.

   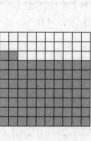

   Part B: Write seventy-two out of a hundred as a decimal.

   0.72

   Part C: Write seventy-two out of a hundred as a fraction in simplest form.

   $\frac{18}{25}$

31. Asia spends three hours biking. How many minutes does she bike? Write an equation.

   180 minutes; $60 \times 3 = 180$

Page 203 • At the Farm

Task Scenario		
Students will use a data set to create a line plot, analyze data from the line plot, subtract fractions, convert decimals to fractions, and compare decimals.		
Depth of Knowledge		DOK1, DOK2, DOK3
Part	**Maximum Points**	**Scoring Rubric**
A	4	Full Credit: 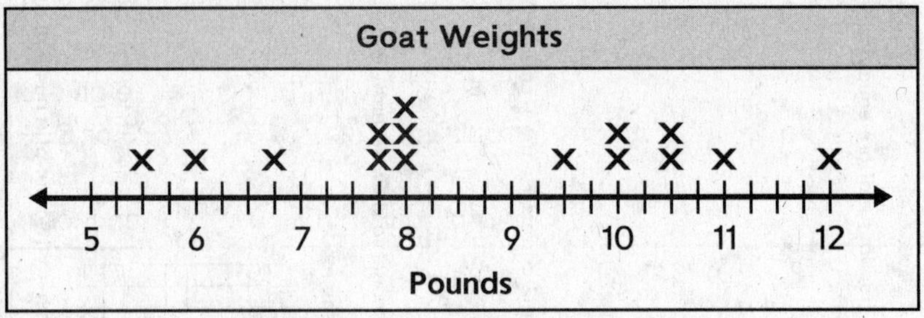 The student correctly creates a line plot from the data given, including all labels for the number line and a title. Partial Credit (3 points) will be given for creating a line plot from the data given but omitting the labels for the number line and title **OR** correctly creating the line plot including labels and title, but incorrectly plotting one to five data points. Partial Credit (2 points) will be given for creating a line plot with 10–14 of the data points plotted correctly but omitting the labels for the number line and title **OR** correctly creating the line plot including labels and title and correctly plotting 6–10 data points. Partial Credit (1 point) will be given for creating a line plot with 6–10 of the data points plotted correctly but omitting the labels for the number line and title **OR** correctly creating the line plot including labels and title and correctly graphing 0–5 data points. No credit will be given for an incorrect answer.
B	3	Full credit: The student correctly identifies the weight of the heaviest goat (12 lb.), the weight of the lightest goat ($5\frac{1}{2}$ lb.), and the highest frequency measurement (8 lb). Partial Credit (1 point) will be given for each correct answer. No credit will be given for three incorrect answers.

Benchmark 4 Performance Task Rubric, continued

Part	Maximum Points	Scoring Rubric
C	2	**Full Credit:** $12 - 5\frac{1}{2} = 6\frac{1}{2}$ Partial credit (1 point) will be given for correctly identifying the difference between the weights of the heaviest and lightest goats without a correct explanation. No credit will be given for an incorrect answer.
D	2	**Full Credit:** The student correctly converts 8.35 to $8\frac{35}{100}$ or $\frac{835}{100}$ or $8\frac{7}{20}$ and correctly locates it on a number line. Partial Credit (1 point) will be given if the student correctly converts 8.35 to $8\frac{35}{100}$ or $\frac{835}{100}$ or $8\frac{7}{20}$ **OR** correctly locates it on a number line. No credit will be given for an incorrect answer.
E	3	**Full Credit:** The student correctly locates 8.35 and 8.58 on the number line and correctly writes the comparison 8.35 < 8.58. Partial credit (2 points) will be given for correctly locating one number on the number line and correctly writing the comparison **OR** correctly locating both numbers on the number line but omitting or incorrectly writing the comparison. Partial credit (1 point) will be given for correctly locating one number on the number line but omitting or incorrectly writing the comparison **OR** for omitting or incorrectly placing the numbers on the number line but correctly writing the comparison. No credit will be given for an incorrect answer.
TOTAL	14	